Recognizing Outstanding Ph.D. Research

Aims and Scope

The series "Springer Theses" brings together a selection of the very best Ph.D. theses from around the world and across the physical sciences. Nominated and endorsed by two recognized specialists, each published volume has been selected for its scientific excellence and the high impact of its contents for the pertinent field of research. For greater accessibility to non-specialists, the published versions include an extended introduction, as well as a foreword by the student's supervisor explaining the special relevance of the work for the field. As a whole, the series will provide a valuable resource both for newcomers to the research fields described, and for other scientists seeking detailed background information on special questions. Finally, it provides an accredited documentation of the valuable contributions made by today's younger generation of scientists.

Theses are accepted into the series by invited nomination only and must fulfill all of the following criteria

- They must be written in good English.
- The topic should fall within the confines of Chemistry, Physics, Earth Sciences, Engineering and related interdisciplinary fields such as Materials, Nanoscience, Chemical Engineering, Complex Systems and Biophysics.
- The work reported in the thesis must represent a significant scientific advance.
- If the thesis includes previously published material, permission to reproduce this must be gained from the respective copyright holder.
- They must have been examined and passed during the 12 months prior to nomination.
- Each thesis should include a foreword by the supervisor outlining the significance of its content.
- The theses should have a clearly defined structure including an introduction accessible to scientists not expert in that particular field.

Gregor Rossmanith

Non-linear Data Analysis on the Sphere

The Quest for Anomalies in the Cosmic Microwave Background

Doctoral Thesis accepted by
the Ludwig-Maximilian University of Munich, Germany

 Springer

Author
Dr. Gregor Rossmanith
Max-Planck-Institute
  for Extraterrestrial Physics
Garching
Germany

Supervisor
Prof. Dr. Dr. h.c. Gregor Morfill
Max-Planck-Institute
  for Extraterrestrial Physics
Garching
Germany

ISSN 2190-5053   ISSN 2190-5061   (electronic)
ISBN 978-3-319-03337-2   ISBN 978-3-319-00309-2   (eBook)
DOI 10.1007/978-3-319-00309-2
Springer Cham Heidelberg New York Dordrecht London

Printed on acid-free paper

Springer is part of Springer Science+Business Media (www.springer.com)

Supervisor's Foreword

The analysis of the cosmic microwave background (CMB) radiation has led to major breakthrough discoveries in Cosmology. The mere discovery of this relic radiation in 1964 by Penzias and Wilson (Nobel Prize in 1978) originating from a hot and dense era of the Universe in the very distant past became one of the first and major proofs for the standard Big Bang cosmological model. The first full-sky observations of the CMB with the COBE satellite in the 1990s of the last century (Nobel Prize for Mather and Smoot in 2006) revealed the amazing homogeneity of this radiation—though the shape of detected fluctuations in the temperature (being as small as of the order of $O(10^{-5})$ °C) could corroborate the assumption of an inflationary period right after the Big Bang.

Higher precision measurements of the CMB as performed nowadays with, e.g., the WMAP and PLANCK satellite have helped to make Cosmology a precision science and to further constrain inflationary scenarios by investigating possible higher order correlations, i.e., non-Gaussianities (NGs), in the field of temperature and polarization fluctuations.

Testing for higher order (nonlinear) correlations has also become a rich field of research in the analysis of complex (chaotic) systems, where it became immediately obvious that a linear description of the complex dynamics of still comparably simple deterministic chaotic systems is already insufficient. Rather, one had to find new ways to describe the behavior of the system. One very successful approach turned out to be the quantification of the phase space portrait, e.g., the strange attractor, formed by the trajectory of the system.

The work presented by Dr. Rossmanith deals with the adaptation and application of two key concepts of complex systems theory, namely estimators for local scaling properties and the method of surrogates, for the search of non-Gaussianities in the CMB.

For calculating so-called scaling indices as estimators for the scaling properties of a CMB map, the idea of embedding a data set in a higher dimensional artificial representation space has been employed. This technique has its origins in Whitneys, Takens, and Sauers embedding theorems, which state that the topology of the phase space structure of a dynamical system is preserved when its dynamics (e.g.

the attractor) is reconstructed in an artificial embedding space by using the method of delay-coordinates. Applying these ideas less formally to image processing tasks allows for the combination of the spatial and intensity (or color) information of the image pixels by representing these physically diverse quantities as points in a common embedding space. The definition of highly sensitive nonlinear local filters relying on the calculation of the local scaling of the point set is thus made possible.

Mostly, tests for NG in Cosmology are performed using models for either Gaussian density fluctuations or special types of non-Gaussianity like NGs of the local type, which can be deduced from special shapes of the potential of the inflaton field.

To develop complementary and model-independent tests for NGs that are able to investigate any deviation from Gaussianity, the method of surrogates was adapted and applied to the case of CMB data analysis. The basic idea of this approach—first having been presented in the seminal paper by Theiler et al., 1992—is to wipe out the higher order correlations by shuffling or replacing the Fourier phases, while exactly preserving the linear properties, i.e., the power spectrum. A comparison of the original data with its surrogates—here using scaling indices as test statistics—then reveals whether higher order correlations have led to phase correlations that were then destroyed or not. Since the phase shuffling can also be restricted to previously selected ranges of scales, the development for dedicated tests for scale-dependent NGs is made possible for the first time.

Dr. Rossmanith performed for the first time a dedicated band-wise analysis with scaling indices using the five-year WMAP data. He found significant signatures for NGs and asymmetries. Further, no dependences on the chosen frequency band were detected, which suggests that the signatures cannot be attributed to yet not understood foreground effects.

The applicability of the method of surrogates for CMB analyses was demonstrated for the largest scales using foreground-reduced full-sky maps derived from the WMAP five-year data release. Highly significant signatures for both non-Gaussianities and asymmetries were once again found. In fact, these detected scale-dependent NGs represent the most significant detection of NGs in the WMAP data to date.

It is also desirable to be able to use the surrogate approach for a cut sky, where the Galactic plane with possible foreground residuals is masked. However, when applying a sky cut, orthonormality of the spherical harmonics no longer holds on this new incomplete sky—making a naive phase shuffling impossible. On the other hand, one can transform the spherical harmonics into a new set of harmonics, which forms an orthonormal basis on the incomplete sky, where phase manipulation can then take place again.

This thesis combines the transformation of the spherical harmonics into a new set of harmonics with the idea of phase shuffling, thus enabling investigations by means of surrogates on an incomplete sky. The feasibility of this approach was once again first demonstrated for the case of large-scale NGs, which also allows for a direct comparison of the full sky and cut sky results.

A scaling index analysis then showed strong non-Gaussianities and pronounced asymmetries, which are consistent with the full sky results and persist even when removing larger parts of the sky. This result confirms that the influence of the Galactic plane is not responsible for these deviations from Gaussianity and isotropy. In the absence of an explanation in terms of Galactic foregrounds or known systematic artifacts, the signatures must so far be taken to be cosmological at high significance. These findings would strongly disagree with predictions of isotropic cosmologies with single-field slow-roll inflation and might even point to a violation of the Cosmological principle.

Of course, those results need further confirmation. Currently, the much more precise CMB data taken with the PLANCK satellite are analyzed with the methodologies outlined in this thesis and results about the large-scale anomalies are expected in the very near future.

A more detailed investigation of the presence and absence of phase correlations as identified in observational CMB data may, on the other hand, also shed more light on the meaning of Fourier phases for complex structures in general. Those findings are of great interest in many (interdisciplinary) fields of research—namely whenever the Fourier representation of sufficiently complex data sets plays a crucial role in the representation and analysis of the data.

Madrid, February 2013   Prof. Dr. Dr. h.c. Gregor Morfill

Reference

1. J. Theiler, S. Eubank, A. Longtin, B. Galdrikian, J.D. Farmer, Physica D **58**, 77 (1992)

Preface

One of the key challenges in Cosmology today is to probe both statistical isotropy and Gaussianity of the primordial density perturbations, which are imprinted in the cosmic microwave background (CMB) radiation. While single-field slow-roll inflation predicts the CMB to fulfill these two characteristics, more complex models may give rise to anisotropy and/or non-Gaussianity. A detection or non-detection allows therefore to discriminate between different models of inflation and significantly improves the understanding of the basic conditions of the very early Universe.

In this work, a detailed CMB non-Gaussianity and isotropy analysis of the five- and seven-year observations of the WMAP satellite is presented. On the one hand, these investigations are performed by comparing the data set with simulations, which is the usual approach for these kinds of analyses. On the other hand, a new model-independent approach is developed and applied in this work. Starting from the random phase hypothesis, so-called surrogate maps are created by shuffling the Fourier phases of the original maps for a chosen scale interval. Any disagreement between the data and these surrogates points towards phase correlations in the original map, and therefore—if systematics and foregrounds can be ruled out—towards a violation of single-field slow-roll inflation.

The construction of surrogate maps only works for an orthonormal set of Fourier functions on the sphere, which is provided by the spherical harmonics exclusively on a complete sky. For this reason, the surrogate approach is for the first time combined with a transformation of the full sky spherical harmonics to a cut sky version. Both the single surrogate approach as well as the combination with the cut sky transformation are tested thoroughly to assess and then rule out the effects of systematics. Thus, this work not only represents a detailed CMB analysis, but also provides a completely new method to test for scale-dependent higher order correlations in complete or partial spherical data sets, which can be applied in different fields of research.

In detail, the applications of the above methods involve the following analyses: First, a detailed study of several frequency bands of the WMAP five-year data release is accomplished by means of a scaling index analysis, whereby the data are compared to simulations. Special attention is paid to anomalous local features, and ways to overcome the problem of boundary effects when excluding foreground-influenced parts of the sky. After this, the surrogate approach is for the first time applied to real CMB data sets. In doing so, several foreground-reduced full sky maps from both the five- and seven-year WMAP observations are used. The analysis includes different scale intervals and a huge amount of checks on possible systematics. Then, another step forward is taken by applying the surrogate approach for the first time to incomplete data sets, again from the WMAP five- and seven-year releases. The Galactic Plane, which is responsible for the largest amount of foreground contribution, is removed by means of several cuts of different sizes. In addition, different techniques for the basis transformation are used.

In all of these investigations, remarkable non-Gaussianities and deviations from statistical isotropy are identified. In fact, the surrogate approach shows by far the most significant detection of non-Gaussianity to date. The band-wise analysis shows consistent results for all frequency bands. Despite a thorough search, no candidate for foreground or systematic influences could be found. Therefore, the findings of these analyses have so far to be taken as cosmological, and point on the one hand towards a strong violation of single-field slow-roll inflation, and question on the other hand the concept of statistical isotropy in general.

Future analyses of the more precise measurements of the forthcoming PLANCK satellite will yield more information about the origin of the detected anomalies.

Contents

Chapter 1
Introduction and Theoretical Background

1.1 A Short Sketch of the Standard Model of Cosmology

The recent years are often referred to as the "Golden Age of Cosmology": Many new experiments, carried out with telescopes on the ground, with balloon missions in the sky, or with satellites and probes in space, lead to a wide range of new results at an unprecedented accuracy. It was possible to establish what is referred to as the "Standard Model of Cosmology" (see e.g. [1–3]): The birth of the Universe occurring by means of a hot *Big Bang*, followed by an extremely short time period called *inflation*, in which the Universe grows by an unbelievable factor between 10^{30} and 10^{50} during less than 10^{-30} s. A few minutes after inflation, primordial *nucleosynthesis* takes place, where the combination of protons and neutrons creates the first nuclei, that again unite with electrons to form atoms around 400,000 years later in the process of *recombination*. From this time on, radiation can travel nearly freely through space due to much less free electrons that could scatter it, and therefore it forms what we call today the *cosmic microwave background* (CMB). The Big Bang Cosmology hypothesises a flat, homogeneous and isotropic universe, which is represented in the Friedmann-Lemaître-Robertson-Walker metric. The development of structure formation is described by the ΛCDM-*Model*. It assumes the Universe to be filled with cold dark matter, and makes use of the cosmological constant Λ. This constant is often referred to as some kind of "vacuum energy", and is responsible for the accelerated expansion of our Universe today.

This combination of theories offers indeed an elegant explanation to a multiplicity of observations. For example, the observed linear correlation between the redshift and distance of supernovae is a strong indication of an expanding universe [4–7]. This relation is today represented by means of the Hubble-Parameter $H(t)$. Furthermore, the measured abundance of light elements, that is ^1H, ^2H, ^3He, ^4He and ^7Li, in the Universe can be successfully explained by nucleosynthesis [8–12]. But one of the strongest arguments for the Standard Model is the measurement of the CMB radiation. Not only the detection itself, but also the fact that its spectrum describes a nearly perfect black body supports the whole concept of modern Cosmology (e.g. [13]).

G. Rossmanith, *Non-linear Data Analysis on the Sphere*, Springer Theses, DOI: 10.1007/978-3-319-00309-2_1, © Springer International Publishing Switzerland 2013

Despite this good agreement to many observations, some parts of the Standard Model are not yet confirmed by detections. Two well-known examples are the question of the existence of Dark Matter as well as Dark Energy, and—after a positive answer—its nature and origin. This open issue leaves some space for alternative ideas (e.g. modified gravity [14, 15] instead of these two quantities, or the local void model [16–18] as an alternative to Dark Energy). Moreover, some results of cosmological analyses seem to disagree to some of the above mentioned theories. For instance, a couple of investigations question the Gaussianity of the CMB and with it the present model of inflation. This challenge is in fact one of the most interesting issues of today's Cosmology. In general, one is bound to say that the Standard Model still offers various open questions. However, the strongly increasing amount of new experiments and analyses as well as the number of people involved shows that we reached a period of high-precision Cosmology like never before. Therefore, we might be able to solve the majority of these open questions in the near future.

This thesis carries out two main tasks: On the one hand, this work accomplishes a detailed search for anisotropies and non-Gaussianities in the CMB. On the other hand, a completely new model-independent approach of data analysis is developed, which allows to test for scale-dependent higher order correlations in complete as well as partial spherical data sets. This method is—for the first time—applied to CMB data, although it can also be used beyond the scope of Cosmology.

The outline of this work is as following: In this Chapter, we will give a short review about the theoretical background of CMB investigations, including the important role of inflation, the basic characteristics of the microwave background itself, and an overview of the current status of CMB observations. The methods and statistics, that form the basis of the CMB analysis in this thesis, will be outlined in detail in Chap. 2. In particular, the attention will be drawn towards the scaling index statistics, the fundamentals of the method of surrogates as well as the construction of an orthonormal basis for a cut sky analysis. Chapter 3 will inform about the measurements and observational difficulties of the WMAP satellite, whose data will be used throughout the whole work. In the following Chap. 4, an analysis by means of the scaling index method examing the WMAP five-year data will be performed. The implementation of the surrogate approach, in combination with the scaling index method, will be accomplished in Chaps. 5 and 6. The analyses include different scale intervals and a large amount of checks on possible systematics. The results will be given for the WMAP five- as well as seven-year data. In Chap. 7, the surrogate approach is for the first time applied to incomplete WMAP data sets. In doing so, the Galactic Plane, which is responsible for the largest amount of foreground contribution, is removed by means of several sky cuts of different sizes. In addition, different techniques for the construction of an orthonormal basis on these partial data sets are used. Finally, we conclude in Chap. 8.

1.2 The Role of Inflation

1.2.1 Inflation as an Improvement of the Standard Model

Inflation was introduced to remove some of the weaknesses that Big Bang theory did suffer from, whereupon the most important one is the horizon problem. In the following, we will give a short overview on every problem and the way inflation solves it (cf. [19–22]).

The Horizon Problem

The distance that light can cover since the Big Bang ($t = 0$) up to today ($t = t_0$) is expressed by the *comoving horizon*

$$\tau = \int_0^{t_0} a(t)^{-1} dt \, ,$$

which also sets the constraints for causal contact between particles. Here, $a(t)$ describes the scale factor. In combination with the Hubble-Parameter $H(t)$, this formula can as well be expressed by means of the *comoving Hubble radius* $(a(t)H(t))^{-1}$:

$$\tau = \int_0^{a_0} \frac{da}{H(t)a^2(t)} = \int_0^{a_0} d \ln a (a(t)H(t))^{-1}$$

Without inflation, the comoving Hubble radius increases monotonically ([21, 22]), hence leading to a growing τ. In this case, CMB radiation, which travels through space nearly since the Big Bang, could not be in causal contact if arriving at the earth from two opposite directions. In contrary to that, the measurements of the microwave background show that it is very close to perfect isotropy (e.g. [23], see also the following Chapter). These measurements are very difficult to explain if there was no causal contact that could bring the radiation into equilibrium. This challenge is known as the *horizon problem*.

Inflation offers a solution to that problem, since it describes a short period, in which the comoving Hubble radius drastically shrinks with time. In other words, the Hubble scale remains constant while the Universe grows dramatically. The *observable Universe*, being consistent with the comoving horizon, is suddenly completely contained in a blown up region that has been very small before inflation occurred. In this small region, causal contact was of course taking place, which offers an elegant explanation for the isotropy of the CMB.

The Flatness Problem

The Friedmann equations rank among the most important equations of Cosmology since they describe the fundamental issue of expansion (or contraction) of the Universe. The first of the two equations,

$$\left(\frac{\dot{a}}{a}\right)^2 = \frac{8\pi G}{3}\rho - \frac{k}{a^2} + \frac{\Lambda}{3} ,$$

where G, ρ, k, a and Λ refer to the gravitational constant, the energy density, the curvature parameter, the scale factor and the Cosmological Constant, can also be written including the density parameters

$$\Omega = \frac{\rho}{\rho_c} \quad , \quad \Omega_\Lambda = \frac{\Lambda}{3H^2} ,$$

at which

$$\rho_c = \frac{3H^2}{8\pi G}$$

describes the critical density. Note that both density parameters are time dependent, though this notation is left out for this Chapter due to simplicity reasons. After some simple algebraic transformations, we obtain [22]

$$|\Omega + \Omega_\Lambda - 1| = \left|\frac{k}{(aH)^2}\right| . \tag{1.1}$$

This equation accurately identifies the departure of the Universe from flatness, $\Omega + \Omega_\Lambda = 1$, and shows that only the term on the right side is responsible for any deviation. Taking a closer look reveals that, once again, the comoving Hubble radius appears in this equation. As stated above, without inflation the expression $(aH)^{-1}$ would monotonically grow with time, and therefore also any departure $|\Omega + \Omega_\Lambda - 1|$. It turns out that the setting $\Omega + \Omega_\Lambda = 1$ marks an *unstable fixed point* [22], meaning that any non-flatness of the Universe increases. Turning this argument upside down, we obtain the fact that any deviation from flatness today must have been even smaller in the past. The constraints are extremely tight, e.g. for the time of nucleosynthesis we obtain an upper limit of 10^{-16} [22]. This strict limitation of selectable curvature settings is called the *flatness problem*.

With the implementation of inflation, the comoving Hubble radius $(aH)^{-1}$ decreases during this time period and the above constraints do not hold anymore. In fact, the Universe is significantly moved towards flatness, and the value $\Omega + \Omega_\Lambda = 1$ now describes an *attractor* [22]. A multiplicity of curvature values is appropriate again, thereby solving the flatness problem.

The Abundances of Magnetic Monopoles

Models of the *Grand Unified Theory* (GUT) predict the creation of nonrelativistic magnetic monopoles in great number, caused by spontaneous symmetry breaking at an energy scale of about 10^{16} GeV [19, 20]. Up to today, these (hypothetical) particles could not be detected [24–26], putting a challenge to the Big Bang model.

Again, this problem can be easily solved by introducing inflation. The rapid expansion of the Universe lowers the frequency density of the magnetic multipoles significantly, and brings theory and observation together.

Structure Formation

In addition to a solution for the problems from above, inflation offers an elegant explanation for the formation of the large scale structure in the Universe today [21, 22]. During the early Universe, microscopic density perturbations appear, generated by quantum fluctuations of the inflaton field, which is the scalar field responsible for the inflationary expansion, as we will see in the next section. Since the Universe grows dramatically during the time of inflation, the size of these density perturbations increases to large scales. Through gravitational instability, more and more mass accumulates in the regions of higher density. This leads at first to single stars and in the end to a distribution of galaxy clusters, and therefore to a structure formation just as we observe it today. The density perturbations are reflected in the CMB (see also Sect. 1.3.2), which is therefore often referred to as the *seed* of the large scale structures.

In the standard single-field slow-roll inflation, the primordial perturbations are assumed to be Gaussian and scale-invariant, which implies a nearly scale free power spectrum $P(k)$ (see also below) that completely describes these pertubations. However, this implies that no higher-order correlations exist and the phases of the Fourier coefficients are random. In fact, the latter assumption will be the main interest of this work, and will be discussed below in more detail.

1.2.2 Basic Concept of Inflation

Inflation describes an extremely short time period taking place at around 10^{-34} s after the Big Bang, in which the Universe is thought to expand by a factor between 10^{30} and 10^{50} at an energy scale of around 10^{15} GeV (for reviews see e.g. [19–22]). It is part of the theories characterising the very early Universe, and has still to be taken as speculative. But, as shown above, it matches perfectly to a lot of observations and removes some of the major weaknesses of Big Bang theory.

The inflation scenario was first proposed in [27] and pursued in [28] and [29] to today's standard concept, which is often denoted as *single-field slow-roll* inflation. However, there exists a large number of different inflation theories nowadays, often including multiple fields: inflation based on supersymmetry, superstring and supergravity models, F-Term or D-Term inflation, brane model inflation, curvaton scenarios, warm inflation, DBI inflation, and many more (for an overview see e.g. [30–32] and references therein). In this Chapter, we will only focus on the classical design of [28, 29], because it is still the most accepted one in Cosmology today and its basic concepts described here already form a groundwork for many of the alternative inflation theories. According to the usual notation in Cosmology, we define the speed of light equal to unity: $c = 1$.

A starting point for the inflation scenario is the request for a shrinking comoving Hubble radius $(a(t)H(t))^{-1}$:

$$\frac{d}{dt}(a(t)H(t))^{-1} < 0 \tag{1.2}$$

where

$$H(t) = \frac{\dot{a}(t)}{a(t)}$$

represents the Hubble-Parameter. As we have seen in the previous section, this condition is the bottom line for solving a couple of main problems of the original Big Bang theory. It can be shown with the help of the Einstein equations that requirement (1.2) corresponds to an accelerated expansion realized by an universe with negative pressure p [22]:

$$\frac{d}{dt}(a(t)H(t))^{-1} < 0 \quad \Rightarrow \quad \frac{d^2a(t)}{dt} > 0 \quad \Rightarrow \quad p > -\frac{\rho}{3}$$

Hereby, ρ denotes the energy density. It is possible to construct such a situation by implementing a scalar field ϕ, called the *inflaton field*, whose dynamics is described by the action

$$S = \int d^4x \sqrt{-\det(g_{\mu\nu})} \left(\frac{1}{2}R + \frac{1}{2}g^{\mu\nu}\partial_\mu\phi\partial_\nu\phi - V(\phi)\right)$$

with R, $g^{\mu\nu}$ and $V(\phi)$ corresponding to the Ricci Scalar, the metric tensor and the potential of the scalar field ϕ. The latter is illustrated in Fig. 1.1. Two so called *slow-roll parameters*

$$\epsilon = -\frac{\dot{H}}{H^2} \quad , \quad \nu = -\frac{\ddot{\phi}}{H\dot{\phi}}$$

set constraints to inflation: The accelerated expansion only takes place if the condition $\epsilon < 1$ holds, while the requirement $|\nu| < 1$ provides the acceleration to retain for a sufficiently long period [20–22].

This generation of a time period with rapid expansion, by introducing the inflaton scalar field, is the basic framework of inflation. The episode of inflation takes place

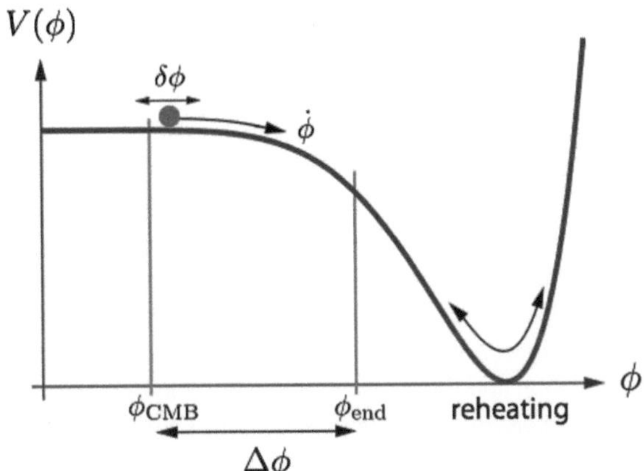

Fig. 1.1 The potential $V(\phi)$ of the inflaton field ϕ. Inflation takes place in the flat part of the potential. At ϕ_{CMB}, the CMB fluctuations are created, while at ϕ_{end}, inflation ends when the kinetic energy equals the potential energy in the steeper part of $V(\phi)$. After that, the scalar is oscillating around the minimum of the potential. Figure taken from [22]

during the time the scalar is located in the flat part of the potential $V(\phi)$. It ends when the field enters the steeper part of the potential and oscillates around the minimum.

An important aspect of this framework is the connection to the characteristics of the CMB (see also Sect. 1.4): It is possible to set a relation between the slow-roll parameters ϵ, ν and the amount of non-Gaussianity in the primordial fluctuations [33]. It was shown in [33–37] that non-Gaussianities are highly suppressed by the slow-roll parameters in the standard model of inflation. Therefore, if the model accurately represents the true conditions of the early Universe, there should be no detectable primordial non-Gaussianity in the CMB. The question if the current measurements agree with the last statement is one of the main topics in this work.

1.3 The Cosmic Microwave Background Radiation

1.3.1 Origin

Shortly after inflation, the Universe was still extremely hot, but was cooling down with time because of its expansion. At a time of approximately $t = 380\,000$ years, the temperature dropped to $T \approx 3000\,\mathrm{K}$, which was cool enough for atomic nuclei and electrons to unite and form the first helium and hydrogen atoms [21, 23], which is referred to as *recombination*. This process can be described formally by means of the Saha equation, which puts the fraction of ionized atoms X—here for hydrogen only—into a relation with temperature [1]:

$$\frac{1-X}{X^2} = \frac{4\sqrt{2}\zeta(3)}{\sqrt{\pi}}\eta \left(\frac{T}{m_e}\right)^{\frac{3}{2}} \exp\left(\frac{13.6\,eV}{T}\right)$$

Hereby, $13.6\,eV$ is the ionization energy of hydrogen. The parameters η and m_e represent the ratio of baryons to photons and the electron mass.

Before recombination, the free electrons were preventing the existing photons to travel freely by means of a rapid Thomson scattering [20]. In other words, the mean free path of a photon was much smaller than the Hubble length H_0^{-1} [21], which expresses the distance a photon would have to travel to reach an observer today. After recombination, the amount of free electrons dropped significantly, leading to a mean free path much longer than the Hubble length [21]. From this time period on, the photons travel nearly freely from every point in space and into every direction. This radiation is termed *Cosmic Microwave Background* (CMB) or *surface of last scattering*, and the moment when the photons were not scattered anymore is referred to as *decoupling* or *time of last scattering*.

1.3.2 Characteristics

The COBE satellite, launched in November 1989, provided for the first time a full-sky measurement of the CMB anisotropies (for the corresponding figure see p. xxx). It also measured for the first time the frequency spectrum of the microwave background radiation. The observation revealed the CMB spectrum to be a nearly perfect *black body* [13, 38]. The remarkably high agreement is illustrated in Fig. 1.2. One can draw the conclusion that before recombination, the Universe was in thermal equilibrium due to the rapid collisions of photons with free electrons, since under these conditions, the frequency spectrum of electromagnetic radiation is represented by the one of a black body [20]. This was maintained in the photons also during and after recombination.

While the CMB has a high energy in the beginning, it features today a temperature of only $2.725\,K$ [40]. This seems to be an odd fact at first sight, since we stated the photons to travel nearly freely through space. But it can easily be explained with the help of its characteristics noticed above: Since it represents a black body, the temperature of the radiation is inverse proportional to the wavelength λ at the peak of the spectrum [41]:

$$T \propto \lambda^{-1}$$

This wavelength λ of the photons is in turn directly proportional to the scale factor, since it is naturally part of the Universe and gets affected by the expansion as well [20],

$$\lambda \propto a(t).$$

In summary, we obtain a direct relation between the temperature ratio and the scale factor ratio from today t_0 and the time of decoupling t_*:

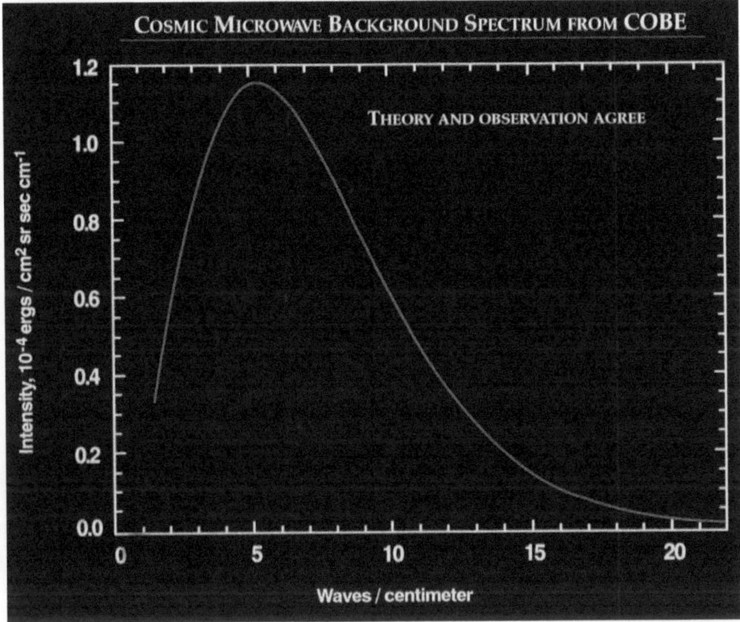

Fig. 1.2 The CMB frequency spectrum as measured by the COBE satellite. Uncertainties are only a small fraction of the line thickness. The spectrum represents a nearly perfect black body. Figure taken from [39] with acknowledgements to the COBE team

$$\frac{T(t_*)}{T(t_0)} = \frac{a(t_0)}{a(t_*)}$$

To express spatial (from the observer) as well as chronological (from today) distances in Cosmology, the *redshift* z is a commonly used parameter. It can be defined via the scale factor [20] as

$$z = \frac{a(t_0)}{a(t)} - 1.$$

With the help of the above mentioned dependencies, the redshift of decoupling z_* can be obtained by using the calculated temperature of the CMB during this time period and the observed temperature of the CMB today [21]:

$$z_* = \frac{T(t_*)}{T(t_0)} - 1$$

Latest observations determine this parameter to $z_* \approx 1090$ [23].

Since the process of decoupling happened in every point in space at the same time, the CMB is said to be highly isotropic. Observations confirm these considerations:

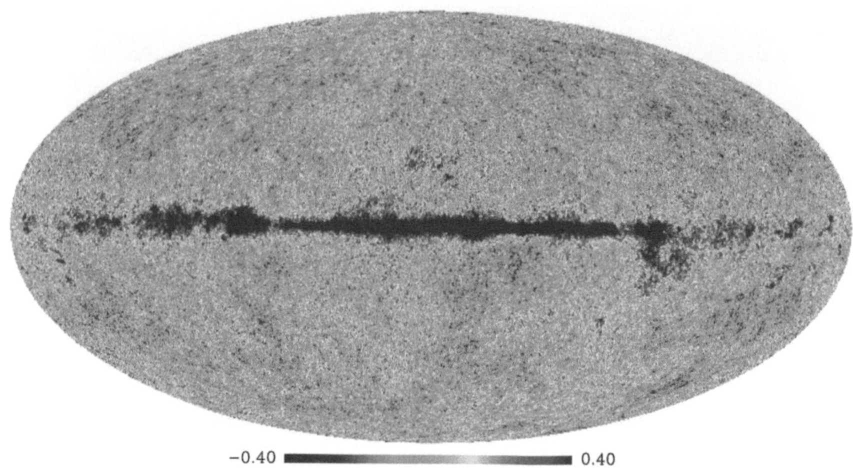

Fig. 1.3 CMB anisotropy ΔT in mK as measured by the WMAP team [42]. The map shows the V-band of the first year results, and is arranged in galactic coordinates in form of a mollweide projection. The *red region* in the centre corresponds to the galactic centre, which strongly distorts the measurements of the background radiation

The temperature fluctuations ΔT of the measured radiation are around five magnitudes smaller than its mean [42],

$$\frac{\Delta T}{T} \approx 10^{-5}.$$

However, this only holds if one has subtracted the dipole from the CMB measurements. This dipole is induced by the Earth's motion with reference to the microwave background and is around a hundred times larger than the usual temperature fluctuations [21],

$$\frac{\Delta T_{Dipole}}{T} \approx 10^{-3}.$$

Figure 1.3 presents the CMB anisotropies ΔT in form of a full sky map as measured by the WMAP probe after the first year observations. Except for the part in the centre, the temperature seems indeed to be highly isotropic at first sight. However, small anisotropies can be measured, which represent the seed points of structure formation, as already mentioned above. These anisotropies are mainly a consequence of the tiny density fluctuations during the time of recombination, whose gravitational potentials the photons have to overcome [22].

The very high temperatures in the centre of Fig. 1.3 are due to foreground contaminations caused by our galaxy. In Sect. 3.2, we will examine the technical difficulties of the observations of the microwave background in more detail.

In contrast to that any other intrinsic anisotropy of the CMB would be of immense cosmological interest [21, 43–45] and is also the main subject throughout this work. We will discuss the basic principles of anisotropies of the CMB in more detail in Sect. 1.4.

1.3.3 Notations

There is more than one possibility of expressing the CMB anisotropies in a formal way. On the one hand, we can describe the temperature fluctuations by writing $\Delta T(\vec{x}) = T(\vec{x}) - \langle T(\vec{x})\rangle$ as above, here with $\vec{x}$ representing the direction in which the temperature is measured. As we will see in Sect. 3.1, the WMAP data is available as a pixelised sky, $\Delta T(x_i)$, $i = 1, ..., N_{pix}$, where N_{pix} denotes the pixel number.

On the other hand, one can remember the fact that the photons reach the observer from every direction, therefore can the microwave background be seen as the surface of a sphere S. Hence, the temperature map $\Delta T(\vec{x})$ can as well be expressed via the spherical harmonics $Y_{\ell m} : S \to \mathbb{C}$ and their coefficients $a_{\ell,m}$ with $\ell \geq 0$, $-\ell \leq m \leq \ell$, [20–22]:

$$T(\vec{x}) = \sum_{\ell=0}^{\infty} \sum_{m=-\ell}^{\ell} a_{\ell m} Y_{\ell m}(\vec{x}) \tag{1.3}$$

In addition, the direction vector $\vec{x}$ is often replaced by a combination of latitude and longitude, $\vec{x} = (\theta, \varphi)$. The harmonics are the spherical analogue of a Fourier series. The first sum in (1.3) theoretically ranges to infinity, but it is usual to set a maximum $\ell \leq \ell_{max}$. Hence, we obtain $(\ell_{max} + 1)^2$ different harmonics $Y_{\ell m}$ in total.

The $Y_{\ell m}$ represent a set of orthonormal, complex valued basis functions on the sphere (see e.g. [41, 46]). They are defined via the Legendre polynomials $P_\ell^m(x)$ as

$$Y_{\ell m}(\theta, \varphi) = \sqrt{\frac{2\ell + 1}{4\pi} \frac{(\ell - m)!}{(\ell + m)!}}\, P_\ell^m(\cos\theta)\, e^{im\varphi}$$

with

$$P_\ell^m(x) = \frac{(-1)^m}{2^\ell \ell!}(1 - x^2)^{m/2} \frac{d^{\ell+m}}{dx^{\ell+m}}(x^2 - 1)^\ell$$

The index m specifies the angular orientation of the spherical harmonic, while the index ℓ is responsible for the characteristic angular size and is therefore isotropic [47].

Each $Y_{\ell m}$ comes up with a parameter $a_{\ell m}$, which can be calculated as an integral over the complete sphere S, [22]

$$a_{\ell m} = \int_S \overline{Y}_{\ell m}(\vec{x})\, T(\vec{x})\, d\Omega\,, \tag{1.4}$$

with $\overline{Y}_{\ell m}$ denoting the complex conjugate of $Y_{\ell m}$. The parameters $a_{\ell m}$ are complex valued as well, and can be written as $a_{\ell m} = |a_{\ell m}| \, e^{i\phi_{\ell m}}$ with an amplitude $|a_{\ell m}|$ and a phase $\phi_{\ell m}$. Since the resulting map $T(\vec{x})$ only contains real values, it holds:

$$Y_{\ell,m} = (-1)^{|m|} \overline{Y}_{\ell,-m}(x) \tag{1.5}$$

which, in combination with (1.4), also leads to

$$a_{\ell,m} = (-1)^{|m|} \overline{a}_{\ell,-m}(x) \, . \tag{1.6}$$

Disregarding the simplification due to ℓ_{max}, the set of $a_{\ell m}$'s contains the entire information of the temperature fluctuation map $T(\vec{x})$.

Another very important quantity in this context is the *power spectrum*,

$$C_\ell = \frac{1}{2\ell + 1} \sum_m |a_{\ell m}|^2 \, , \tag{1.7}$$

which is a sum of the amplitudes of the Fourier coefficients, and describes the Fourier transform of the two-point correlation function. The underlying idea of the power spectrum is the above mentioned assumption that the microwave background radiation is isotropic. Thus, the distribution of the $a_{\ell m}$'s should be independent of the index m. This assumption is reflected in Eq. (1.7).

The power spectrum is a very useful tool in Cosmology and offers a multitude of applications. The main reason for this is the fact that if the CMB is Gaussian, the power spectrum completely characterises all the information of the temperature anisotropies $T(\vec{x})$ (see the following Sect. 1.4). This would of course represent an immense simplification: The entire information about the structural properties of the CMB map would be compressed in only ℓ_{max} values. However, the Gaussianity of the CMB is controversial and its investigation the main topic of this work.

In addition, the shape of the power spectrum is connected to the physics of the beginnings of our Universe. In the existing baryon-photon plasma, the gravitational attraction of and the radiation repulsion in the density enhanced regions acted together to produce acoustic oscillations [21, 43]. These created temperature fluctuations in the CMB, which are again reflected as a wave-shaped profile in the power spectrum. The most recent WMAP data detect three peaks at $\ell \approx 200, 550, 800$, whereupon the first is clearly the most pronounced [48]. By measuring the shape and the location of these peaks, a plenitude of conclusions for cosmological parameters can be drawn [44, 45, 47]. This is mostly done by assuming a flat adiabatic ΛCDM model, and adjusting its parameters in a way that the resulting power spectrum agrees with the measurements [49]. In doing so, constraints can be drawn about e.g. the cosmic baryon and matter densities Ω_b and Ω_m, the age t_0 and the curvature k of the Universe, the epoch of matter-radiation equality z_{eq} and the primordial helium mass fraction Y_{He} [48, 49].

1.4 The Challenge of Anomalies in the CMB

One of the key questions in Cosmology today is the question, if the measured microwave background satisfies the requirements that the theoretical framework is demanding for: *Isotropy* and *Gaussianity*, both already mentioned above. Although these two properties often appear in combination, they describe different effects:

- *(Statistical) Isotropy* demands that there is no preferred direction with particular structural characteristics. In combination with the Copernican Principle—presuming that our spatial position in space is not exceptional—this would lead to homogeneity (see e.g. [50]). One would speak of *anisotropy*, if there exists at least one direction with significant deviations from the usual structural behaviour.
- *Gaussianity* refers in this context to the assumption, that the coefficients of the spherical harmonics are independent Gaussian random variables [44],

$$P(a_{\ell m})\, da_{\ell m} = \frac{1}{\sqrt{2\pi C_\ell}} \exp(-\frac{a_{\ell m}^2}{2C_\ell})\, da_{\ell m} \ .$$

The variance is expressed by the respective value of the power spectrum. Therefore, if Gaussianity holds, the power spectrum characterises all the information that is contained in the CMB. From the equation it follows that the amplitudes $|a_{\ell m}|$ ought to be Rayleigh distributed, while the phases $\phi_{\ell m}$ follow a uniform distribution in the interval $[-\pi, \pi]$. In contrast to that, there is no specific definition for *non-Gaussianity*, except for the request for a deviation from Gaussianity in any possible way.

As already mentioned in Sect. 1.2.2, both characteristics are a consequence of the physics of single-field slow-roll inflation [21, 22, 37, 44, 45, 51–54]. However, other inflationary theories [33, 34, 55–68] could induce anisotropies or non-Gaussianities. Also, models leading to a scale-dependent non-Gaussianity are conceivable [69, 70]. Besides, there could exist topological defects like cosmic strings [71–74] or particular phenomenons as the occurrence of large voids [75–79], that could generate deviations from the above statements. Thus, a detection or non-detection of anisotropies and non-Gaussianities is of highest interest, since it allows to discriminate between different models of inflation and sheds light on basic conditions of the Universe.

However, there are also some known effects that have influence on Gaussianity and isotropy [21, 43–45, 47, 53]. Those can be roughly divided into effects concerning the physics of the very early Universe—and therefore the primordial CMB—and interactions with the microwave background during the flight of the photons; so-called secondary anisotropies. Both types induce anisotropies, whereupon only the latter can significantly influence Gaussianity as well [43].

The primordial CMB is affected by the *Sachs-Wolfe effect* [80], which describes the already above stated existence of density fluctuations in the early Universe, whose different potential wells the photons have to escape. In addition, *silk damping* [81] takes place since recombination is not happening instantaneously.

In contrast to that, there are the effects on the CMB after decoupling, like e.g. the *integrated Sachs-Wolfe effect* (ISW) [80], which refers to the time-dependent gravitational fields. Their potential wells decrease due to the expansion of the Universe, leading to a blueshift in the photons that pass through these fields. In some literature, early ISW ranks among the primordial effects instead of the secondary anisotropies. Another effect is the *Sunyaev-Zel'dovich* (SZ) effect [82, 83], which can be divided into a *thermal* part, that denotes the Compton scattering of the CMB photons by hot electron gas in galaxy clusters, and a *kinetic* part, which describes diverse scattering due to the motion of the gas in these clusters. As a result of gravitational interactions with matter, *gravitational lensing* [84] is an important part of the secondary anisotropies as well.

Apart from all those effects, there are foregrounds and other technical difficulties distorting the CMB signal and therefore maybe biasing the measured isotropy and Gaussianity. These will be discussed in more detail in Sect. 3.2.

Nowadays, it is still accepted by a large part of the cosmological community that the CMB is both isotropic and non-Gaussian. As stated above, the measured fluctuations $\Delta T(\vec{x})$ of the background radiation seem to agree with the theoretical prediction at first sight. However, there is also a growing number of analyses that nevertheless detect inconsistencies in the data. In the following, we will point out some of these inconsistencies. The methods that were used to analyse the CMB, as well as the basic principles and problems of a non-Gaussianity analysis will then be discussed in more detail in Chap. 2.

There is a plethora of analyses that discovered deviations from Gaussianity in a general sense, as for example [85–104].

But there are also some more specific anomalous features in the CMB, that were subject to a lot of investigations. The most important ones are the alignment of the large multipoles, the power asymmetry of the temperature fluctuations, and the Cold Spot. We will discuss these in more detail in the following.

The detection of an *anomalous alignment* of the quadrupole ($\ell = 2$) and the octopole ($\ell = 3$) was first reported in [108] and [109]. In [110], this alignment was found to involve the entire multipole range $\ell = 2 - 5$, and is since then a topic of various investigations [105, 111–123]. The upper left plot of Fig. 1.4 shows the temperature anisotropies of the quadrupole and the octopole of the WMAP seven-year measurements in combination with the ecliptic plane. The probability for the obvious alignment to happen by chance is around 0.1 % [105]. In addition, the ecliptic plane seems to be correlated to this alignment, and separates a hot spot in the northern sky and a cold spot in the south. Up to today, there is no explanation for this correlation [105].

Another anomalous feature that puts isotropy into question is the discovery of *power asymmetries* in the CMB [106, 124–141]. In the upper right plot of Fig. 1.4, the results of an investigation of a combination of two bands of the WMAP seven-year data is presented. Several separate sets of 100 multipole blocks inside the interval $\ell \in [2, 600]$ were analysed individually. The directions of the dipoles of each of these multipole blocks are indicated by the coloured discs. Obviously, all of these lie very close to each other, but also close to the southern ecliptic pole and to the dipole of the full interval $\ell \in [2, 600]$, that are indicated by a cross and the white hexagon,

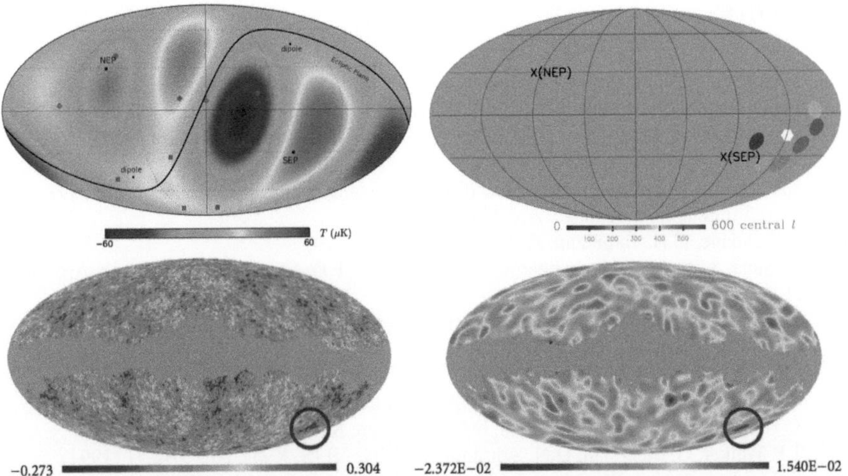

Fig. 1.4 Three different anomalous and currently still unexplained features in the CMB that were detected in the recent years: On the *upper left*, the anomalous alignment of the quadrupole and the octopole that is shown as a plot of the temperature anisotropies of only these two multipoles. The *black curve* marks the position of the ecliptic plane. The projection on the *upper right* refers to the power asymmetry, which is illustrated by the directions of the dipoles of the estimated power distributions when considering blocks of 100 multipoles each. The colours of the discs specify the centres of the multipole ranges, while the *white hexagon* indicates the dipole direction of the full interval $\ell \in [2, 600]$, and the crosses mark the northern and southern ecliptic poles, respectively. In the *lower column*, the Cold Spot is shown as it appears in the temperature map of the WMAP seven-year release (*left*) and in the response map of the applied wavelet analysis (*right*). Figures taken from [105–107]

respectively. This implies that the hemisphere centred in these directions contains more power than the opposite hemisphere. In addition, the different multipoles seem to be correlated to each other. No known systematic effects or foregrounds are found to be able to explain this asymmetry [106].

Finally, *local features*—first and foremost the famous *Cold Spot* [130]—were detected and confirmed by several analyses [77, 78, 133, 136, 142–153]. The original detection of the Cold Spot was accomplished by means of a wavelet analysis of the CMB temperature anisotropies, and is shown in the lower column of Fig. 1.4. Especially in the wavelet response map on the right, the Cold Spot is clearly visible. Systematics or foreground effects were ruled out to be responsible for this local feature, and its chance to happen accidentally is around or less then 1 %, depending on the type of the analysis [107].

To which extent all these analysis are significant is still subject to discussion, though [154–156].

References

1. E.W. Kolb, M.S. Turner, *The Early Universe (Frontiers in Physics)* (Perseus Books / Westview Press, New York, 1994)
2. P.J.E. Peebles, *Principles of Physical Cosmology* (Princeton University Press, Princeton, 1993)
3. J.A. Peacock, *Cosmological Physics* (Cambridge University Press, Cambridge, 1999)
4. G.A. Tammann, A. Sandage, ApJ **151**, 825 (1968)
5. A. Sandage, G.A. Tammann, ApJ **194**, 559 (1974)
6. A. Saha, A. Sandage, L. Labhardt, G.A. Tammann, F.D. Macchetto, N. Panagia, ApJ **486**, 1 (1997)
7. A. Riess, L. Macri, S. Casertano, H. Lampeitl, H.C. Ferguson, A.V. Filippenko, S.W. Jha, W. Li, R. Chornock, ApJ **730**, 119 (2011)
8. P.J.E. Peebles, ApJ **146**, 542 (1966)
9. R.V. Wagoner, W.A. Fowler, F. Hoyle, ApJ **148**, 3 (1967)
10. J. Yang, M.S. Turner, G. Steigman, D.N. Schramm, K. Olive, ApJ **281**, 493 (1984)
11. T.P. Walker, G. Steigman, D.N. Schramm, K. Olive, H.-S. Kang, ApJ **376**, 51 (1991)
12. B. Fields, S. Sarkar, Phys. Lett. **b592**, 1 (2004)
13. D.J. Fixsen, E.S. Cheng, J.M. Gales, J.C. Mather, R.A. Shafer, E.L. Wright, ApJ **473**, 576 (1996)
14. S. Nojiri, S.D. Odintsov, Phys. Lett. B **576**, 5 (2003)
15. S.M. Carroll, V. Duvvuri, M. Trodden, M.S. Turner, Phys. Rev. D **70**, 043528 (2004)
16. J.W. Moffat, D.C. Tataraski, ApJ **453**, 17 (1995)
17. K. Tomita, ApJ **529**, 26 (2000)
18. M.N. Celerier, Astron. Astrophys. **353**, 63 (2000)
19. V. Mukhanov, *Physical Foundations of Cosmology* (Cambridge University Press, Cambridge, 2005)
20. S. Weinberg, *Cosmology* (Oxford University Press, London, 2008)
21. W.H. Kinney, *Lectures from the Theoretical Advanced Study Institute at University of Colorado, Boulder*, arXiv:0902.1529v2 (2008)
22. D. Baumann, *Lectures from the Theoretical Advanced Study Institute at University of Colorado, Boulder*, arXiv:0907.5424v1 (2009)
23. N. Jarosik, C.L. Bennett, J. Dunkley, B. Gold, M.R. Greason, M. Halpern, R.S. Hill, G. Hinshaw, A. Kogut, E. Komatsu, D. Larson, M. Limon, S.S. Meyer, M.R. Nolta, N. Odegard, L. Page, K.M. Smith, D.N. Spergel, G.S. Tucker, J.L. Weiland, E. Wollack, E.L. Wright, ApJS **192**, 14 (2011)
24. K.A. Milton, Rept. Prog. Phys. **69**, 1637 (2006)
25. H. Kadowaki, N. Doi, Y. Aoki, Y. Tabata, T.J. Sato, J.W. Lynn, K. Matsuhira, Z. Hiroi, J. Phys. Soc. Jpn. **78**, 103706 (2009)
26. M. Detrixhe, D. Besson, P.W. Gorham, P. Allison, B. Baughmann, J.J. Beatty, K. Belov, S. Bevan, W.R. Binns, C. Chen, P. Chen, J.M. Clem, A. Connolly, D. DeMarco, P.F. Dowkontt, M.A. Duvernois, C. Frankenfeld, E.W. Grashorn, D.P. Hogan, N. Griffith, B. Hill, S. Hoover, M.H. Israel, A. Javaid, K.M. Liewer, S. Matsuno, B.C. Mercurio, C. Miki, M. Mottram, J. Nam, R.J. Nichol, K. Palladino, A. Romero-Wolf, L. Ruckman, D. Saltzberg, D. Seckel, G.S. Varner, A.G. Vieregg, Y. Wang, Phys. Rev. D **83**, 023513 (2011)
27. A.H. Guth, Phys. Rev. D **23**, 347 (1981)
28. A. Linde, Phys. Lett. **108B**, 389 (1982)
29. A. Albrecht, P. Steinhardt, Phys. Rev. Lett. **48**, 1220 (1982)
30. A. Linde, *Particle Physics and Inflationary Cosmology* (Harwood, Chur, 1990)
31. D. Lyth, A. Riotto, Phys. Rept. **314**, 1 (1999)
32. A.R. Liddle, D.H. Lyth, *Cosmological Inflation and Large-Scale Structure* (Cambridge University Press, Cambridge, 2000)
33. V. Acquaviva, N. Bartolo, S. Matarrese, A. Riotto, Nucl. Phys. B **667**, 119 (2003)

34. T.J. Allen, B. Grinstein, M.B. Wise, Phys. Lett. B **197**, 66 (1987)
35. T. Falk, R. Rangarajan, M. Srednicki, Phys. Rev. D **46**, 4232 (1992)
36. A. Gangui, F. Lucchin, S. Matarrese, S. Mollerach, ApJ **430**, 447 (1994)
37. J.M. Maldacena, JHEP **05**, 013 (2003)
38. J.C. Mather, E.S. Cheng, D.A. Cottingham, R.E. Eplee, D.J. Fixsen, T. Hewagama, R.B. Isaacman, K.A. Jensen, S.S. Meyer, P.D. Noerdlinger, S.M. Read, L.P. Rosen, R.A. Shafer, E.L. Wright, C.L. Bennett, N.W. Boggess, M.G. Hauser, T. Kelsall, S.H. Moseley, R.F. Silverberg, G.F. Smoot, R. Weiss, D.T. Wilkinson, APJ **420**, 439 (1994)
39. http://lambda.gsfc.nasa.gov/
40. B. Gold, N. Odegard, J.L. Weiland, R.S. Hill, A. Kogut, C.L. Bennett, G. Hinshaw, X. Chen, J. Dunkley, M. Halpern, N. Jarosik, E. Komatsu, D. Larson, M. Limon, S.S. Meyer, M.R. Nolta, L. Page, K.M. Smith, D.N. Spergel, G.S. Tucker, E. Wollack, E.L. Wright, ApJS **192**, 15 (2011)
41. D. Meschede, *Gerthsen Physik* (Springer Verlag, Heidelberg, 2010)
42. C.L. Bennett, M. Halpern, G. Hinshaw, N. Jarosik, A. Kogut, M. Limon, S.S. Meyer, L. Page, D.N. Spergel, G.S. Tucker, E. Wollack, E.L. Wright, C. Barnes, M.R. Greason, R.S. Hill, E. Komatsu, M.R. Nolta, N. Odegard, H.V. Peirs, L. Verde, J.L. Weiland, ApJS **148**, 1 (2003)
43. K. Nakamura et al., Particle data group. J. Phys. G **37**, 075021 (2010)
44. V. Rubakov, A. Vlasov, Phys. At. Nucl. **75**(9), 1123 (2012)
45. E. Komatsu, K.M. Smith, J. Dunkley, C.L. Bennett, B. Gold, G. Hinshaw, N. Jarosik, D. Larson, M.R. Nolta, L. Page, D.N. Spergel, M. Halpern, R.S. Hill, A. Kogut, M. Limon, S.S. Meyer, N. Odegard, G.S. Tucker, J.L. Weiland, E. Wollack, E.L. Wright, ApJS **192**, 18 (2011)
46. W. Demtröder, *Experimentalphysik 3* (Springer, Heidelberg, 2004)
47. D. Samtleben, S. Staggs, B. Winstein, Ann. Rev. Nucl. Part. Sci. **57**, 245 (2007)
48. D. Larson, J. Dunkley, G. Hinshaw, E. Komatsu, M.R. Nolta, C.L. Bennett, B. Gold, M. Halpern, R.S. Hill, N. Jarosik, A. Kogut, M. Limon, S.S. Meyer, N. Odegard, L. Page, K.M. Smith, D.N. Spergel, G.S. Tucker, J.L. Weiland, E. Wollack, E.L. Wright, ApJS **192**, 16 (2011)
49. L. Page, M.R. Nolta, C. Barnes, C.L. Bennett, M. Halpern, G. Hinshaw, N. Jarosik, A. Kogut, M. Limon, S.S. Meyer, H.V. Peiris, D.N. Spergel, G.S. Tucker, E. Wollack, E.L. Wright, ApJS **148**, 233 (2003)
50. R. Maartens, http://arxiv.org/pdf/1104.1300v1
51. E. Komatsu, Class. Quantum Gravity **27**, 124010 (2010)
52. D. Jeong, E. Komatsu, ApJ **703**, 1230 (2009)
53. A.P.S. Yadav, B.D. Wandelt, Adv. Astron. **2010**, 565248 (2010)
54. N. Bartolo, E. Komatsu, S. Matarrese, A. Riotto, Phys. Rept. **402**, 103 (2004)
55. A.D. Linde, JETP **40**, 1333 (1984)
56. A.A. Starobinskiǐ, JETP Lett. **42**, 152 (1985)
57. D. Polarski, A.A. Starobinskiǐ, Phys. Rev. D **50**, 6123 (1994)
58. A.H. Linde, V. Mukhanov, Phys. Rev. D **56**, R535 (1997)
59. P.J.E. Peebles, ApJ Lett. **438**, L1 (1997)
60. T. Moroi, T. Takahashi, Phys. Lett. B **522**, 215 (2001)
61. K. Enqvist, M.S. Sloth, Nucl. Phys. **B626**, 395 (2002)
62. N. Bartolo, S. Matarrese, A. Riotto, Phys. Rev. D **65**, 103505 (2002)
63. F. Bernardeau, J.-P. Uzan, Phys. Rev. D **66**, 103506 (2002)
64. D.H. Lyth, D. Wands, Phys. Lett. B **524**, 5 (2002)
65. D.H. Lyth, C. Ungarelli, D. Wands, Phys. Rev. D **67**, 023503 (2003)
66. E. Silverstein, D. Tong, Phys. Rev. D **70**, 103505 (2004)
67. M. Sasaki, J Valiviita, D. Wands, Phys. Rev. D **74**, 103003 (2006)
68. I.G. Moss, C. Xiong, JCAP **0704**, 007 (2007)
69. C. Armendariz-Picon, T. Damour, V. Mukhanov, Phys. Lett. B **458**, 209 (1999)
70. J. Garriga, V. Mukhanov, Phys. Lett. B **458**, 219 (1999)
71. N. Kaiser, A. Stebbins, Nat. **310**, 391 (1984)
72. N. Turok, D. Spergel, Phys. Rev. Lett. **64**, 2736 (1990)
73. N. Turok, ApJ Lett. **473**, L5 (1996)

74. R. Jeannerot, J. Rocher, M. Sakellariadou, Phys. Rev. D **68**, 103514 (2003)
75. K.T. Inoue, J. Silk, ApJ **648**, 23 (2006)
76. M. Cruz, E. Martínes-González, P. Vielva, J.M. Diego, M. Hobson, N. Turok, MNRAS **390**, 913 (2008)
77. V.G. Gurzadyan, A.A. Kocharyan, A&A Lett. **492**, L33 (2008)
78. V.G. Gurzadyan, A.E. Allahverdyan, T. Ghahramanyan, A.L. Kashin, H.G. Khachatryan, A.A. Kocharyan, H. Kuloghlian, S. Mirzoyan, E. Poghosian, G. Yegorian, A&A **497**, 343 (2009)
79. I. Masina, A. Notari, JCAP **09**, 028 (2010)
80. R.K. Sachs, A.M. Wolfe, ApJ **147**, 73 (1967)
81. J. Silk, ApJ **151**, 459 (1968)
82. Y.B. Zel'dovich, R.A. Sunyaev, Ap&SS **4**, 301 (1996)
83. R.A. Sunyaev, Y.B. Zeldovich, Comments on Ap&SS **4**, 173 (1972)
84. A. Einstein, Science **84**, 2188 (1936)
85. E. Komatsu, N. Afshordi, N. Bartolo, D. Baumann, J.R. Bond, E.I. Buchbinder, C.T. Byrnes, X. Chen, D.J.H. Chung, A. Cooray, P. Creminelli, N. Dalal, O. Dore, R. Easther, A.V. Frolov, K.M. Górski, M.G. Jackson, J. Khoury, W.H. Kinney, L. Kofman, K. Koyama, L. Leblond, J.-L. Lehners, J.E. Lidsey, M. Liguori, E.A. Lim, A. Linde, D.H. Lyth, J. Maldacena, S. Matarrese, L. McAllister, P. McDonald, S. Mukohyama, B. Ovrut, H.V. Peiris, C. Raeth, A. Riotto, Y. Rodriguez, M. Sasaki, R. Scoccimarro, D. Seery, E. Sefusatti, U. Seljak, L. Senatore, S. Shandera, E.P.S. Shellard, E. Silverstein, A. Slosar, K.M. Smith, A.A. Starobinsky, P.J. Steinhardt, F. Takahashi, M. Tegmark, A.J. Tolley, L. Verde, B.D. Wandelt, D. Wands, S. Weinberg, M. Wyman, A.P.S. Yadav, M. Zaldarriaga, *Astro2010: The Astronomy and Astrophysics Decadal Survey*, Science White Papers, vol. 158 (2009)
86. L.-Y. Chiang, P.D. Naselsky, O.V. Verkhodanov, M.J. Way, ApJ Lett. **590**, L65 (2003)
87. P.D. Naselsky, A.G. Doroshkevich, O.V. Verkhodanov, ApJ Lett. **599**, L53 (2003)
88. P.D. Naselsky, L.-Y. Chiang, P. Olesen, O.V. Verkhodanov, ApJ **615**, 45 (2004)
89. P.D. Naselsky, L.-Y. Chiang, P. Olesen, I. Novikov, Phys. Rev. D **72**, 063512 (2005)
90. E. Komatsu, A. Kogut, M.R. Nolta, C.L. Bennett, M. Halpern, G. Hinshaw, N. Jarosik, M. Limon, S.S. Meyer, L. Page, D.N. Spergel, G.S. Tucker, L. Verde, E. Wollack, E.L. Wright, ApJS **148**, 119 (2003)
91. P. Coles, P. Dineen, J. Earl, D. Wright, MNRAS **350**, 989 (2004)
92. I.J. O'Dwyer, H.K. Eriksen, B.D. Wandelt, J.B. Jewell, D.L. Larson, K.M. Górski, A.J. Banday, S. Levin, P.B. Lilje, ApJ Lett. **617**, L99 (2004)
93. L.-Y. Chiang (1), P.D. Naselsky, Int. J. Mod. Phys. D **15**, 1283 (2006)
94. L.-Y. Chiang, P.D. Naselsky, P. Coles, ApJ **664**, 8 (2007)
95. L.-Y. Chiang, P.D. Naselsky, MNRAS Lett. **380**, L71 (2007)
96. A.P.S. Yadav, B.D. Wandelt, PRL **100**, 181301 (2008)
97. H.K. Eriksen, G. Huey, A.J. Banday, K.M. Górski, J.B. Jewell, I.J. O'Dwyer, B.D. Wandelt, ApJ **665**, 1L (2007)
98. O. Rudjord, F.K. Hansen, X. Lan, M. Liguori, D. Marinucci, S. Matarrese, ApJ **701**, 369 (2009)
99. K.M. Smith, L. Senatore, M. Zaldarriaga, JCAP **0909**, 006 (2009)
100. O. Rudjord, F.K. Hansen, X. Lan, M. Liguori, D. Marinucci, S. Matarrese, ApJ **708**, 1321 (2010)
101. P. Cabella, D. Pietrobon, M. Veneziani, A. Balbi, R. Crittenden, G. de Gasperis, C. Quercellini, N. Vittorio, MNRAS **405**, 961 (2010)
102. A. Bernui, M.J. Reboucas, Int. J. Mod. Phys. D **19**, 1411 (2010)
103. A. Bernui, M.J. Reboucas, A.F.F. Teixeira, Int. J. Mod. Phys. D **19**, 1405 (2010)
104. A. Bernui, M.J. Reboucas, Phys. Rev. D **81**, 063533 (2010)
105. C.J. Copi, D. Huterer, D.J. Schwarz, G.D. Starkman, Adv. Astron. **2010**, 847541 (2010)
106. F.K. Hansen, A.J. Banday, K.M. Górski, H.K. Eriksen, P.B. Lilje, ApJ **704**, 1448 (2009)
107. P. Vielva, Adv. Astron. **2010**, 592094 (2010)
108. M. Tegmark, A. Oliveira-Costa, A. Hamilton, Phys. Rev. D **68**, 123523 (2003)
109. A. Oliveira-Costa, M. Tegmark, M. Zaldarriaga, A. Hamilton, Phys. Rev. D **69**, 063516 (2004)

110. K. Land, J. Magueijo, Phys. Rev. Lett. **95**, 071301 (2005)
111. D.J. Schwarz, G.D. Starkman, D. Huterer, C.J. Copi, Phys. Rev. Lett. **93**, 221301 (2004)
112. J. Weeks, eprint arXiv:astro-ph/0412231 (2004)
113. J.P. Ralston, Pankaj Jain. Int. J. Mod. Phys. D **13**, 1857 (2004)
114. K. Land, J. Magueijo, Phys. Rev. Lett. **72**, 101302 (2005)
115. C.J. Copi, D. Huterer, D.J. Schwarz, G.D. Starkman, MNRAS **367**, 79 (2006)
116. K. Land, J. Magueijo, MNRAS **367**, 1714 (2006)
117. C.J. Copi, D. Huterer, D.J. Schwarz, G.D. Starkman, Phys. Rev. D **75**, 023507 (2007)
118. K. Land, J. Magueijo, MNRAS **378**, 153 (2007)
119. A. Gruppuso, C. Burigana, JCAP **0908**, 004 (2009)
120. D. Hanson, A. Lewis, Phys. Rev. D **80**, 063004 (2009)
121. M. Frommert, T.A. Ensslin, MNRAS **403**, 1739 (2010)
122. A. Gruppuso, K.M. Górski, JCAP **03**, 019 (2010)
123. D. Sarkar, D. Huterer, C.J. Copi, G.D. Starkman, D.J. Schwarz, Astropart. Phys. **34**, 591 (2011)
124. H.K. Eriksen, F.K. Hansen, A.J. Banday, K.M. Górski, P.B. Lilje, ApJ **605**, 14 (2004)
125. H.K. Eriksen, D.I. Novikov, P.B. Lilje, A.J. Banday, K.M. Górski, ApJ **612**, 64 (2004)
126. H.K. Eriksen, A.J. Banday, K.M. Gorski, F.K. Hansen, P.B. Lilje, ApJ Lett. **660**, L81 (2007)
127. F.K. Hansen, P. Cabella, D. Marinucci, N. Vittorio, ApJ Lett. **607**, L67 (2004)
128. F.K. Hansen, A.J. Banday, K.M. Górski, MNRAS **354**, 641 (2004)
129. C.-G. Park, MNRAS **349**, 313 (2004)
130. P. Vielva, E. Martínes-González, R.B. Barreiro, J.L. Sanz, L. Cayón, ApJ **609**, 22 (2004)
131. K. Land, J. Magueijo, MNRAS **357**, 994 (2005)
132. A. Bonaldi, S. Ricciardi, S. Leach, F. Stivoli, C. Baccigalupi, G. De Zotti, MNRAS **382**, 1791 (2007)
133. C. Räth, P. Schuecker, A.J. Banday, MNRAS **380**, 466 (2007)
134. A. Bernui, Phys. Rev. D **78**, 063531 (2008)
135. A. Bernui, W.S. Hipolito-Ricaldi, MNRAS **389**, 1453 (2008)
136. D. Pietrobon, A. Amblard, A. Balbi, P. Cabella, A. Cooray, D. Marinucci, Phys. Rev. D **78**, 103504 (2008)
137. P. Vielva, J.L. Sanz, MNRAS, arXiv:0910.3196v2
138. J. Hoftuft, H.K. Eriksen, A.J. Banday, K.M. Górski, F.K. Hansen, P.B. Lilje, ApJ **699**, 985 (2009)
139. F. Paci, A. Gruppuso, F. Finelli, P. Cabella, A. De Rosa, N. Mandolesi, P. Natoli, MNRAS **407**, 339 (2010)
140. L. Cayón, MNRAS **405**, 1084 (2010)
141. N.E. Groeneboom, M. Axelsson, D.F. Mota, T. Koivisto, arXiv:1011.5353v1
142. P. Mukherjee, Y. Wang, ApJ **613**, 51 (2004)
143. L. Cayón, L. Jin, A. Treaster, MNRAS **362**, 826 (2005)
144. M. Cruz, E. Martínes-González, P. Vielva, L. Cayón, MNRAS **356**, 29 (2005)
145. M. Cruz, M. Tucci, E. Martínes-González, P. Vielva, MNRAS **369**, 57 (2006)
146. M. Cruz, N. Turok, P. Vielva, E. Martínes-González, M. Hobson, Science **318**, 1612 (2007)
147. M. Cruz, L. Cayón, E. Martínes-González, P. Vielva, J. Jin, ApJ **655**, 11 (2007)
148. J.D. McEwen, M.P. Hobson, A.N. Lasenby, D.J. Mortlock, MNRAS **359**, 1583 (2005)
149. J.D. McEwen, M.P. Hobson, A.N. Lasenby, D.J. Mortlock, MNRAS Lett. **371**, L50 (2006)
150. J.D. McEwen, M.P. Hobson, A.N. Lasenby, D.J. Mortlock, MNRAS **388**, 659 (2008)
151. P. Vielva, Y. Wiaux, E. Martínes-González, P. Vandergheynst, MNRAS **381**, 932 (2007)
152. V.G. Gurzadyan, A.L. Kashin, H.G. Khachatryan, A.A. Kocharyan, E. Poghosian, D. Vetrugno, G. Yegorian, Europhys. Lett. **91**, 19001 (2010)
153. P. Vielva, E. Martínes-González, M. Cruz, R.B. Barreiro, M. Tucci, MNRAS, arXiv:1002.4029v2
154. R. Zhang, D. Huterer, Astropart. Phys. **33**, 69 (2010)
155. E.F. Bunn, Proceedings of Rencontres de Moriond (2010)
156. C.L. Bennett, R.S. Hill, G. Hinshaw, D. Larson, K.M. Smith, J. Dunkley, B. Gold, M. Halpern, N. Jarosik, A. Kogut, E. Komatsu, M. Limon, S.S. Meyer, M.R. Nolta, N. Odegard, L. Page, D.N. Spergel, G.S. Tucker, J.L. Weiland, E. Wollack, E.L. Wright, ApJS **192**, 17 (2011)

Chapter 2
Methods for Testing the Non-Gaussianity of the CMB

2.1 Statistical Tests for Non-Gaussianity

2.1.1 Basic Framework

Cosmologists searching for non-Gaussianity in the CMB have to deal with two major fundamental statistical problems. First, it is not clear what to look for and which way is the best for doing that. Let us recall that non-Gaussianity—which an analyst intends to find or to rule out—can occur in various ways, since it is only defined as anything except Gaussianity (see Sect. 1.4). Therefore, there is a nearly infinite number of thinkable investigations. Besides, any analysis resulting in a non-detection of anomalous behaviour does not prove the CMB to be Gaussian, but just rules out a single type of non-Gaussianity corresponding to the characteristics of the analysis.

However, also the detection of peculiarities in the data does not immediately imply intrinsic non-Gaussianities in the microwave background, because the high amount of foreground contributions could leave hidden imprints in particular in the results of more complex analyses.

The second fundamental statistical problem of the CMB is the fact that there is only one realisation. Irrespective from foregrounds or technical difficulties, there is in theory no way to tell if a possibly detected anomalous behaviour is due to different underlying physics or just a statistical fluke. There is the idea of using the polarisation of the CMB as a new independent sample (e.g. [1–3]), however this strongly depends on the characteristics of the investigation and can not be seen as a solution in general.

The first of the two problems lead to an amazingly large amount of different measures for non-Gaussianity. A short overview over some of these measures will be given in Sect. 2.2.1. The second problem yields the fact that one has to interpret the results of the different analyses with great caution. Since any kind of possible measure is "allowed" to be used, its choice could sometimes be motivated by the characteristics of the data itself. The choice would therefore be an *a posteriori* one (cf. [4, 5]). The fact that all analyses naturally focus on anomalous features in the data, combined with a plethora of different measures used today, could lead to some

G. Rossmanith, *Non-linear Data Analysis on the Sphere*, Springer Theses, DOI: 10.1007/978-3-319-00309-2_2, © Springer International Publishing Switzerland 2013

sort of preselection and therefore lowers the validity of possible anomalies. However, this does not mean that all analyses working on non-Gaussianity become redundant. In fact, every investigation claiming deviations from the theoretical properties of the microwave background is supposed to obtain a very significant result, that is in the best case confirmed by different measures. Apart from that, checks on systematics and ruling out foreground effects as a cause for the deviations is always necessary.

A common technique to search for non-Gaussianity in the microwave background is to construct simulated maps, that are Gaussian random fields which mimic the properties of the ΛCDM model. The analysis is then performed on both the data and a set of these simulations. Eventually, a comparison of the results gives information about how well the measured CMB corresponds to the theoretical demands (e.g. [6–8]). This is also accomplished in this work in Chap. 4 for the WMAP five-year data set. On the other hand, some investigations make use of particular assumptions about the nature of the non-Gaussianities by parametrising it with e.g. the non-linear coupling parameter f_{NL} (e.g. [2, 9, 10], see also below).

Clearly, both procedures depend on the model or the assumptions that are implemented. However, it might be favourable to rely on as few requirements as possible. A complementary and elegant way to investigate the non-Gaussianity of the CMB is an analysis that is completely *model-independent*. In the following Chapter, we will introduce the surrogate method, that describes one possibility of a thorough data-driven, i.e. model-independent investigation.

2.1.2 Surrogates on the Complete Sky

The concept of constructing surrogates from a given data set originates from the field of non-linear time series analysis. The basic idea was introduced in the paper of Theiler et al. [11] and subsequently applied to several different data sets, like fluid convection, sunspots, as well as electroencephalograms [12], and was continuously developed [13, 14]. Further, constrained randomisation has already been used before to generate CMB data sets with random phases as a technique for analysing the effect of cosmic strings. This was combined with a multifractal formalism in [15] for detecting cosmic string induced non-Gaussianity on synthetic CMB data sets. The surrogate method can be applied on nearly all complex systems, as outlined in [16] for the climate, stock-market and the heart-beat variability. In combination with scaling indices, which is the measure used throughout this work and will be presented in Sect. 2.2.2 below, surrogates were applied for large scale structure analysis [17] and non-Gaussianity investigations on simulated two-dimensional temperature maps [18].

The starting point for the surrogates technique is a given data set and some null hypothesis, whose validity in the data set is to be tested. The fundamental idea is then to generate surrogate data sets from the original data, which are consistent with the null hypothesis. Apart from the characteristics that are affected by the hypothesis, these surrogates share exactly the same properties as the original data set. Next, the

data as well as the surrogates are tested by means of some measure that is sensitive
to characteristics, which could be induced by deviations from the null hypothesis.
If different results are obtained for the original and the set of surrogates, the null
hypothesis is rejected. If not, the hypothesis is confirmed.

We apply this basic concept to CMB non-Gaussianity analysis. As null hypothesis,
we take the *random phase hypothesis*, which is the assumption that the phases $\phi_{\ell m}$ of
the spherical harmonic coefficients $a_{\ell m}$ are independent and identically distributed in
terms of a uniform distribution in the interval $[-\pi, \pi]$ (see Sect. 1.4). This assumption
is on the one hand a very important and fundamental statement. Only if the random
phase hypothesis holds, the construction of the power spectrum, which represents a
compression of the information of a complete CMB map with several million data
points to only around one thousand values, is lossless and therefore fully justified
[19]. On the other hand, the statement of random phases is a direct consequence of
the presumed Gaussianity of the CMB. Since the power spectrum only takes into
account the linear correlations in the map, possible higher-order correlations can
only be contained in the phases and the correlations among them. Thus, the presence
of phase correlations would clearly disagree with Gaussianity. Any detection of
inconsistencies between a CMB data set and surrogates, whose phases do not have
any correlations, would therefore directly identify non-Gaussian behaviour of the
CMB data. For this reason, the method of constructing surrogates used in this work
is based on a phase shuffling technique, which destroys possible phase correlations
of the original data set, and which is consistent with the stated null hypothesis.

Phases were already subject to analyses concerning the formation of the large
scale structure in the Universe [20]. Also, a close look at the quadrupole of the CMB
[21] as well as its foregrounds [22, 23] is possible in terms of a phase analysis.
Investigations searching for phase correlations—and therefore non-Gaussianity—of
the CMB were performed in [24–31] (see also Sect. 2.2.1 for a closer look at the
results).

The method for generating surrogates by shuffling the phases is as follows: As
described in detail in Sect. 1.3.3, a temperature map $T(\vec{x})$, $\vec{x} \in S$, on the complete
sphere S can be expressed by means of spherical harmonics $Y_{\ell m}$,

$$T(\vec{x}) = \sum_{\ell=0}^{\ell_{max}} \sum_{m=-\ell}^{\ell} a_{\ell m} Y_{\ell m}(\vec{x}) \,.$$

The sum consists of $i_{max} = (\ell_{max} + 1)^2$ different summands. The coefficients $a_{\ell m}$
are complex-valued, and can therefore be written in polar coordinates,

$$a_{\ell m} = |a_{\ell m}| \, e^{i \phi_{\ell m}} \,,$$

in which the phases $\phi_{\ell m}$ can be computed as

$$\phi_{\ell m} = \arctan \frac{\mathrm{Im}(a_{\ell m})}{\mathrm{Re}(a_{\ell m})} \,.$$

To ensure that possible outliers in the data, which do not follow the assumed probability distributions as given in Sect. 1.4, are not affecting the results of the further process, one has to implement the following preprocessing. However, possible phase correlations are not affected by these two steps.

- First, the temperature values $T(\vec{x})$ are replaced by a Gaussian distribution in a rank-ordered remapping. We use the expressions $T_{old}(\vec{x})$ and $T_{new}(\vec{x})$ for denoting the temperature values before and after the remapping. Formally, we obtain:

$$T_{new}(\vec{x}_i) = D^{(k)}$$

with $D(x) \sim \mathcal{N}(\mu, \sigma)$ for $x = 1, ..., N_{pix}$, and $D^{(1)} < D^{(2)} < ... < D^{(N_{pix})}$. Hereby, μ and σ denote the mean and the standard deviation of $T_{old}(\vec{x})$, respectively, while k characterises the position of $T_{old}(\vec{x}_i)$ in the rank ordering

$$T_{old}^{(1)} < T_{old}^{(2)} < ... < T_{old}^{(N_{pix})} .$$

- A similar rank ordering is applied to the values of the phases $\phi(i) = \phi_{i(\ell,m)}$:

$$\phi_{new}(i) = D^{(k)}$$

with $D(x) \sim \mathcal{U}([-\pi, \pi])$ for $x = 1, ..., i_{max}$, and $D^{(1)} < D^{(2)} < ... < D^{(i_{max})}$. Similar to above, k describes the position of $\phi_{old}(i)$ in the rank ordering

$$\phi_{old}^{(1)} < \phi_{old}^{(2)} < ... < \phi_{old}^{(i_{max})} .$$

Hence, all detected deviations between the underlying map and the constructed surrogates can only be due to possible phase correlations inside the original data set.

To perform the surrogates method, one has at first to choose a shuffling interval $[\ell_1, \ell_2]$ containing the scales that are of interest in the analysis. This interval may be chosen arbitrarily inside the possible range of all multipoles, $[0, \ell_{max}]$. However, since in data maps for CMB investigations, the monopole and dipole are often subtracted, the lower bound should in this case be larger than one, $\ell_1 \geq 2$. After a convenient interval was chosen, one applies two shuffling steps onto the underlying data map, to generate two kinds of surrogates:

The first step is a shuffling of the phases of all coefficients $a_{\ell m}$, $m > 0$, outside the range $[\ell_1, \ell_2]$. In doing so, all phase correlations that correspond to scales that are not of interest, are destroyed. The resulting map with the shuffled phases is termed *first order surrogate*. The second step is to shuffle the phases of the first order surrogate inside the range of interest $[\ell_1, \ell_2]$, to create a map with no phase correlations at all. This step is to be performed multiple times to obtain several data sets. The resulting maps are denoted *second order surrogates*. In Fig. 2.1, the two phase shuffling steps are illustrated on a ℓ-m-diagram. In each step of this process, the phases with a negative m-value, $\phi_{\ell m}$, $m < 0$, have to be shuffled in the same way as the corresponding phases with the positive m-value, since otherwise the imaginary

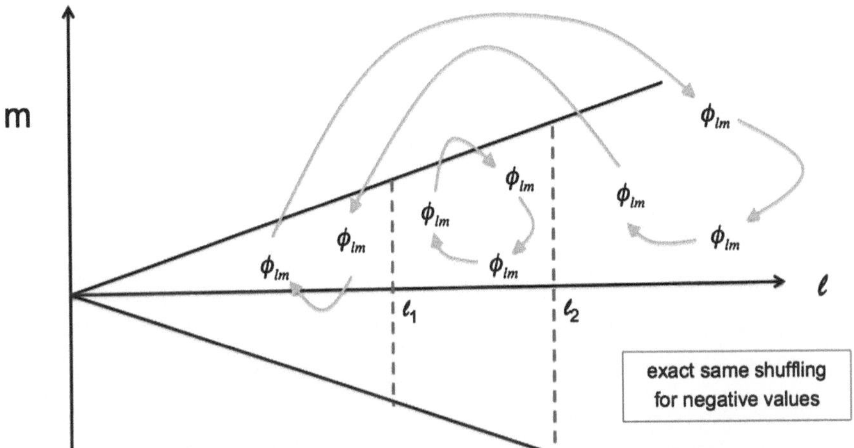

Fig. 2.1 A ℓ-m-diagram for illustration purposes of the two phase shuffling steps: At first, the phases outside $[\ell_1, \ell_2]$ are shuffled to obtain the first order surrogate, and afterwards, the phases inside the interval are shuffled multiple times to create several second order surrogates

parts of the coefficients $a_{\ell m}$ and spherical harmonics $Y_{\ell m}$ would not cancel each other. Note that all surrogates possess by definition exactly the same linear properties, that is the power spectrum, as the underlying map, since the amplitudes $|a_{\ell m}|$ were left unchanged.

The first order surrogate is then compared with the set of second order ones by means of some measure (see the following Sect. 2.2 for a overview of currently used measures in the field of CMB non-Gaussianity). Since the preprocessing steps from above ensure the correct distributions for the temperature values and the phases, any detected discrepancies have to be traced back to the phase correlations inside the first order surrogate, and are therefore a sign of non-Gaussianity *inside the chosen multipole range* of the initial map. Hence, the surrogate method presents a technique to search for deviations from Gaussianity in a range of scales which can be chosen arbitrarily.

A special case of the shuffling technique occurs if one chooses the range $[\ell_1, \ell_2] = [0, \ell_{max}]$ (or $[\ell_1, \ell_2] = [2, \ell_{max}]$ in case of a mono- and dipole reduced map, see above). Since this interval covers the complete range of available multipoles, generating a first order surrogate becomes dispensable. In this case, a comparison between the original map and the second order surrogates shows deviations from Gaussianity on all scales.

In Fig. 2.2, first and second order surrogates of the seven-year ILC map are illustrated with an underlying scale range of interest $[\ell_1, \ell_2] = [20, 60]$.

Despite the advantage of analyses on all arbitrary scales, the surrogates method also possesses a disadvantage: It requires a complete sphere, to ensure that the spherical harmonics are orthogonal. Otherwise, the phases of the underlying map would be coupled, which leads to induced phase correlations. For this reason, the surrogates method is performed on maps, where the foreground influences—especially those

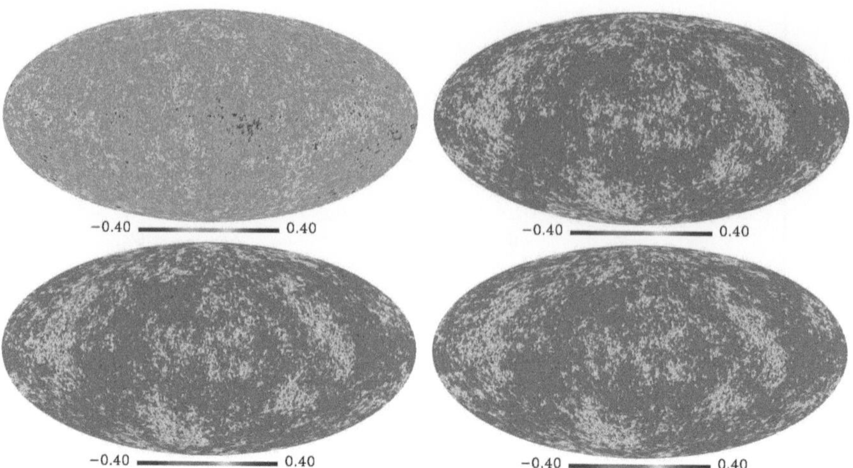

Fig. 2.2 An example of the phase shuffling method: The phases of the underlying seven-year ILC map (*upper left*) are shuffled outside the interval $[\ell_1, \ell_2] = [20, 60]$ to obtain the first order surrogate (*upper right*). Two realisations of an additional shuffling of the phases inside the interval leads to the two second order surrogates (*lower row*). Note that the structural behaviour of the large scales differs between the maps in the *upper row*, but is similar for all the first as well as the second order surrogates

due to the Galactic plane—are reduced to a minimum. This is provided by the ILC or the NILC maps (see Sect. 3.2.2). The results of these investigations are presented in Chaps. 5 and 6. But to accomplish an even more thorough analysis, it is better to mask out highly foreground affected regions like the Galactic plane, which hence puts a problem to the method. However, as a main part of this work, new ways to construct a new set of orthogonal harmonics for incomplete skies were developed, thus solving this problem. These techniques will be presented in detail in the following section, and are applied to data sets in Chap. 7.

2.1.3 Surrogates on an Incomplete Sky

The spherical harmonics form an orthonormal basis set on the complete sphere S. This statement is expressed formally by the equation

$$\int_S Y_{\ell m}(\vec{x})\, Y_{\ell' m'}(\vec{x})\, d\Omega = \begin{cases} 1 & \text{for } \ell = \ell' \text{ and } m = m' \\ 0 & \text{else} \end{cases} \tag{2.1}$$

where $Y_{\ell m}, Y_{\ell' m'}$ characterise two harmonic functions with $\ell, \ell' \geq 0, -\ell \leq m \leq \ell, -\ell' \leq m' \leq \ell'$. This equation describes a fundamental condition. Only if orthogonality holds, the coefficients $a_{\ell m}$ of a map $f(\vec{x})$ are unique.

If one replaces the complete sphere S in Eq. (2.1) by some incomplete sky S^{cut}, the orthogonality of the spherical harmonics obviously vanishes. This leads to severe problems, since in this case the coefficients $a_{\ell m}$ would be coupled. Hence, the random phase hypothesis no longer holds, and the surrogates technique from the previous Sect. 2.1.2 becomes inapplicable.

Incomplete skies often appear in CMB non-Gaussianity analyses: Highly foreground affected regions, in the first place the Galactic plane, strongly influence the Gaussianity of the map. Even the foreground-reduced maps, as provided by the WMAP team, still have obvious artefacts in the Galactic plane (cf. Fig. 3.4 on p. XXX). The best way to deal with this, is to apply a sky cut on these regions (cf. Sect. 3.2). The usage of full-sky maps with minimal Galactic foreground contribution, like the ILC [32] or the NILC map [33], is another solution to the problem, which avoids the sky cut. But the map-making process of these maps could induce phase correlations, which can then not be distinguished from the intrinsic higher-order correlations of the CMB.

However, there are ways to overcome this problem: In [34, 35], a method was presented, which transforms the real-valued spherical harmonics to a new set of harmonics, that is orthonormal on an user-defined cut sky. This method was improved and extended in [36]. In the present work, we adopt these techniques onto the complex-valued spherical harmonics, and combine it with the surrogates analysis, to enable investigations by surrogates on an arbitrary cut sky.

Our goal is to express any CMB temperature map

$$f(\vec{x}) = \sum_{\ell=0}^{\ell_{max}} \sum_{m=-\ell}^{\ell} a_{\ell m} Y_{\ell m}(\vec{x}), \ \vec{x} \in S,$$

on an incomplete sky S^{cut} by means of new coefficients $a_{\ell m}^{cut}$ and new cut sky harmonics $Y_{\ell m}^{cut} : S^{cut} \to \mathbb{C}$,

$$f(\vec{x}) = \sum_{\ell=0}^{\ell_{max}} \sum_{m=-\ell}^{\ell} a_{\ell m}^{cut} Y_{\ell m}^{cut}(\vec{x}), \ \vec{x} \in S^{cut},$$

where $Y_{\ell m}^{cut}$ is an orthogonal basis set on S^{cut}, and thus $a_{\ell m}^{cut}$ being unique.

At first, we write the spherical harmonics and the original coefficients of the underlying map, as well as the harmonics and the coefficients that we would like to obtain, into one vector each. In doing so, we only consider the modes with $m \geq 0$:

$$
\begin{aligned}
Y(\vec{x}) \quad &:= [Y_{0,0}(\vec{x}), Y_{1,0}(\vec{x}), Y_{1,1}(\vec{x}), ..., Y_{\ell_{max},\ell_{max}}(\vec{x})]^T, \\
Y^{cut}(\vec{x}) &:= [Y_{0,0}^{cut}(\vec{x}), Y_{1,0}^{cut}(\vec{x}), Y_{1,1}^{cut}(\vec{x}), ..., Y_{\ell_{max},\ell_{max}}^{cut}(\vec{x})]^T, \\
a \quad &:= [a_{0,0}, a_{1,0}, a_{1,1}, ..., a_{\ell_{max},\ell_{max}}]^T \\
a^{cut} \quad &:= [a_{0,0}^{cut}, a_{1,0}^{cut}, a_{1,1}^{cut}, ..., a_{\ell_{max},\ell_{max}}^{cut}]^T
\end{aligned}
$$

All vectors have the length of $i_{max} := (\ell_{max}+1)(\ell_{max}+2)/2$. With the help of these terms, we can express our goal in a different way: We would like to determine two matrices $B_1, B_2 \in \mathbb{C}^{i_{max} \times i_{max}}$, that transform the vectors of the spherical harmonics and the original coefficients into the cut sky vectors of above, which is characterised formally by the following equations:

$$Y^{cut}(\vec{x}) = B_1\, Y(\vec{x}) \tag{2.2}$$

$$a^{cut} = B_2\, a \tag{2.3}$$

It is possible to evaluate these two matrices by applying several matrix computations onto the vector $Y(\vec{x})$. The first step, important for both the calculation of B_1 and B_2, is the definition of the *coupling matrix* and its cut sky counterpart $C, C^{cut} \in \mathbb{C}^{i_{max} \times i_{max}}$:

$$C \quad := \int_R Y(\vec{x}) Y^*(\vec{x}) d\Omega$$

$$C^{cut} := \int_R Y^{cut}(\vec{x})(Y^{cut})^*(\vec{x}) d\Omega$$

where R characterises an area on the sphere, and Y^* denotes the hermitian transposed of Y. When working with a pixelised sky, like e.g. in the HEALPix environment used for the WMAP data set, one has to replace the integral with a sum over all pixels that belong to R. The coupling matrices can be treated as positive definite for low ℓ_{max}. In addition, C and C^{cut} are hermitian by definition: For the diagonal of C, the components read as $c_{i(\ell,m),i(\ell,m)} = \int_R Y_{\ell m} \overline{Y}_{\ell m} d\Omega$, which is obviously real-valued. Outside the diagonal, we obtain $c_{i(\ell,m),i(\ell',m')} = \int_R Y_{\ell m} \overline{Y}_{\ell'm'} d\Omega = \int_R \overline{\left(\overline{Y}_{\ell m} Y_{\ell'm'}\right)} d\Omega = \overline{c}_{i(\ell',m'),i(\ell,m)}$. The equivalent holds for C^{cut}.

To evaluate B_1, we have to recall the orthonormality condition (2.1) from above. Applying an adequate condition to the incomplete cut sky S^{cut}, it follows that a set of harmonics $Y_{\ell m}^{cut}$, which is orthonormal on the cut sky, needs to fulfil the equation

$$C^{cut} = I_{i_{max}} \; .$$

Hereby, $I_{i_{max}}$ denotes the unit matrix of size i_{max}. We can use Eq. (2.2) to change this condition to

$$B_1 C B_1^* = I_{i_{max}} \tag{2.4}$$

It is possible to decompose the coupling matrix and to calculate a matrix $A \in \mathbb{C}^{i_{max} \times i_{max}}$ which fulfills $C = A A^*$. Hereby, different matrix decomposition methods can be used, as for example the Cholesky or the eigenvalue decomposition. We will discuss this important step of the calculation in more detail below. Applying this decomposition to the above equation leads to

$$(B_1 A)(B_1 A)^* = I_{i_{max}} \; ,$$

which offers the simple solution $B_1 = A^{-1}$. Note that this does not have to be the only possible solution: In general, every matrix B_1 that fulfils Eq. (2.4) is applicable.

For B_2, we rewrite Eq. (1.4), which offered a formula for the computation of $a_{\ell m}$, into a vectorial form,

$$a = \int_S \overline{Y}(\vec{x}) f(\vec{x}) d\Omega$$

or correspondingly

$$a^{cut} = \int_{S^{cut}} \overline{Y}^{cut}(\vec{x}) f(\vec{x}) d\Omega .$$

By inserting (2.2) and replacing the map by means of $f(\vec{x}) = a^T Y(\vec{x})$, we obtain

$$a^{cut} = \int_{S^{cut}} \overline{B}_1 (Y(\vec{x}) Y^*(\vec{x}))^T a \, d\Omega = \overline{B}_1 C^T a .$$

Again, we make use of the above introduced matrix decomposition and apply additionally the result of the first transformation matrix from above, $B_1 = A^{-1}$, which leads to

$$a^{cut} = \overline{B}_1 (AA^*)^T a = A^T a .$$

Thus, it follows $B_2 = A^T$.

So far, we ignored the cut sky harmonics $Y_{\ell m}^{cut}(\vec{x})$ and coefficients $a_{\ell m}^{cut}$ for $m < 0$. For their computation, we make use of Eqs. (1.5) and (1.6). We assume these equations to be valid also in the cut sky regime,

$$Y_{\ell,m}^{cut} = (-1)^{|m|} \overline{Y}_{\ell,-m}^{cut}(x)$$
$$a_{\ell,m}^{cut} = (-1)^{|m|} \overline{a}_{\ell,-m}^{cut}(x) ,$$

and can thus easily get the missing terms. Nevertheless, the above equations could in general lead to a non-orthogonal set of cut sky harmonics, since each $Y_{\ell,-m}^{cut}$ is by definition only orthogonal to its counterpart $Y_{\ell,m}^{cut}$, but possibly not to the rest of the harmonics. Still, for all sky cuts and ℓ-ranges that were used throughout this work, the cut sky harmonics were tested and confirmed to be orthogonal.

In summary, both transformation matrices B_1, B_2 can be easily determined once the decomposition of the coupling matrix $C = AA^*$ was successful, and we obtain

$$Y^{cut}(\vec{x}) = A^{-1} Y(\vec{x}) \tag{2.5}$$

$$a^{cut} = A^T a . \tag{2.6}$$

However, the matrix decomposition is—from a numerical point of view—the most difficult part of the cut sky procedure, since the matrix C grows exponentially with the fourth power of ℓ_{max}. The choice of which decomposition technique one uses has a strong influence on the characteristics of the cut sky harmonics $Y_{\ell m}^{cut}$. In this work, we will apply three different decomposition methods, the *Cholesky*, the *eigenvalue*, and the *singular value decomposition* (cf. e.g. [37]). All three techniques require the coupling matrix to be positive definite, which holds up to some ℓ_{max} that depends on the applied sky cut. The differences between the three decompositions are explained in the following.

Cholesky Decomposition

The easiest approach is the Cholesky decomposition, which was already used for the real-valued cut sky harmonics in [34] and [36]. It defines the matrix A to be lower triangular (and therefore A^* to be upper triangular), and calculates then step by step a solution for each row of A. For example, the first three diagonal elements of A have to fulfil

$$(c_{11}) = (a_{11})^2$$
$$(c_{22}) = (a_{21})^2 + (a_{22})^2$$
$$(c_{33}) = (a_{31})^2 + (a_{32})^2 + (a_{33})^2 \,, \; ...$$

which can be solved in combination with similar (but more complex) equations for the off-diagonal terms. The Cholesky decomposition is implemented in nearly every mathematical software today and provides the fastest results of all three matrix decompositions used in this work. Another advantage is the fact that A is lower triangular. Having a look a Eq. (2.6), one can see that this leads to a comfortable situation: In this case, all cut sky coefficients $a^{cut}_{i^*(\ell,m)}$ only depend on the full sky coefficients of higher multipoles, $a_{i(\ell,m)}, i \geq i^*$. Therefore, a monopole and dipole reduction is still possible, since these are only contained in the first four cut sky coefficients.

Eigenvalue Decomposition

Another possibility is to apply the eigenvalue decomposition, which was also used in [36] (identified there as "singular value decomposition", which is not necessarily wrong, as we will see below). The basic idea relies on the possibility to rewrite the coupling matrix in the following way [37]:

$$C = VWV^* \,,$$

where the columns of V contain the eigenvectors of C, and the diagonal matrix W contains the corresponding eigenvalues. For a positive definite and hermitian C, these eigenvalues are real-valued and larger than zero, and can therefore be used to divide the above term into

$$C = VW^{1/2}(VW^{1/2})^* \,,$$

which leads to the solution $A = VW^{1/2}$. Hereby, $W^{1/2}$ denotes the matrix that contains the square roots of the elements of W.

Singular Value Decomposition

A method very similar to the eigenvalue decomposition is the singular value decomposition. This is based on the fact that one can write [37]

$$C = UWV^* .$$

Hereby, in contrast to above, U contains the eigenvectors of CC^*, V the eigenvectors of C^*C, and the diagonal matrix W the eigenvalues of either CC^* or C^*C, which leads to the same result. These eigenvalues are also termed *singular values* of the matrix C. Since C is hermitian, it follows $CC^* = C^*C$ and therefore $U = V$. Hence, we obtain

$$C = UWU^* = (UW^{1/2})(UW^{1/2})^*$$

and thus the result $A = UW^{1/2}$. When applying this method, it is important to consider the following: For a hermitian matrix like the used coupling matrix C, it can be shown that the resulting matrices A of eigenvector and singular value decomposition are theoretically *exactly* consistent with each other (cf. e.g. [37]). However, this does not hold in practice: The decomposition by means of the singular values yields numerically far better results than the eigenvector decomposition, since it can be applied onto larger coupling matrices (and therefore higher ℓ_{max}) and provides a faster calculation.

There exist two technical procedures that improve the decomposition processes from above:

First, the Cholesky decomposition has the advantage of a triangular transformation matrix. This does not hold for the other two decompositions, but in this case it is again possible to decompose the matrix A into a triangular matrix A' and an unitary matrix U with the same size each,

$$A = A'U ,$$

by applying a *Householder transformation*. The unitary matrix can then be ignored, since it does not change the decomposition equation $C = AA^*$, and therefore one can use A' instead of A. See the Appendix for a more detailed description of this technique.

Second, when applying a *constant latitude cut*, the majority of the terms of the coupling matrix C becomes trivial. This simplifies its calculation as well as its decomposition. See again the Appendix for more details.

Examples of the new cut sky harmonics for two different constant latitude cuts are illustrated in Fig. 2.3.

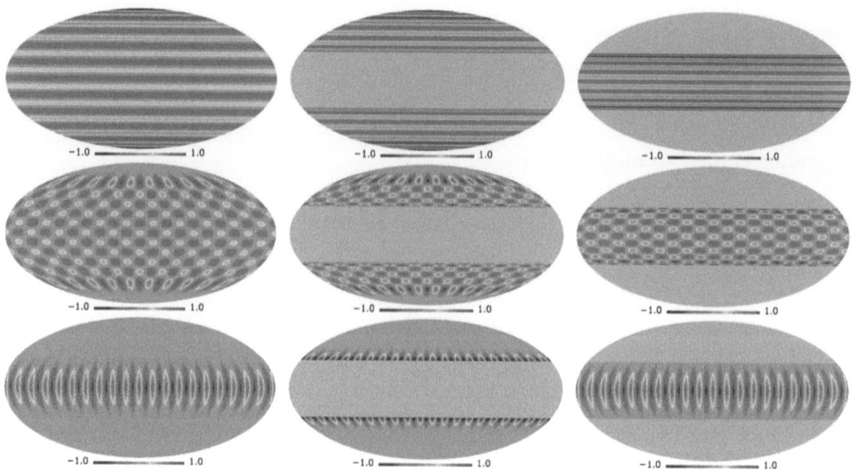

Fig. 2.3 Examples of the original spherical (*first column*) and the new cut sky harmonics for $(\ell, m) = (20, 0)$, $(20, 10)$ and $(20, 20)$ (from *top to bottom*). The harmonics were constructed by means of the singular value decomposition with additional householder transformation for $l_{max} = 20$ and a constant latitude cut of $|b| < 30°$ (*second column*) and $|b| \geq 30°$ (*third column*), respectively. Only the real part of the complex-valued harmonics is shown in each plot

After the calculation of the new sets of cut sky harmonics $Y_{\ell m}^{cut}(\vec{x})$ and coefficients $a_{\ell m}^{cut} = \left| a_{\ell m}^{cut} \right| e^{i\phi_{\ell m}^{cut}}$ corresponding to the underlying map $f(\vec{x})$, one can finally apply the surrogates method. Similar to the previous Sect. 2.1.2, the phases

$$\phi_{\ell m}^{cut} = \arctan \frac{\mathrm{Im}(a_{\ell m}^{cut})}{\mathrm{Re}(a_{\ell m}^{cut})}$$

are shuffled, while the amplitudes $\left| a_{\ell m}^{cut} \right|$ are preserved. Each shuffling results in a new set of $a_{\ell m}^{cut}$'s, which corresponds to one cut sky surrogate map. Special care has to be taken when choosing the shuffling range $[\ell_1, \ell_2]$, since the scales of the structural behaviour of the map might no longer be preserved in the cut sky coefficients. For low ℓ_{max}, a rough scale similarity still holds, though. An example of a surrogate set of the WMAP seven-year ILC map with a multipole limit of $\ell_{max} = 40$ and a shuffling range of $[\ell_1, \ell_2] = [2, 40]$ is presented in Fig. 2.4.

But this result is not satisfying yet because of one remaining problem: By applying the cut sky transformation, the phases of the underlying data map additionally get correlated due to Eq. (2.6). This effect is shown in Fig. 2.5, which illustrates the results of a naive cut sky analysis of a simulated Gaussian CMB map with independent phases by means of scaling indices (see Sect. 2.2.2 below), for a series of increasing constant latitude cuts, that remove $|b| < 10°$, $|b| < 20°$ and $|b| < 30°$ of the Galactic plane. The colour-coded pixels show the σ-normalised deviations between each hemisphere of the original and the surrogate data sets around this pixel. The details of this investigation are not important for the moment and will be discussed in more detail in Chap. 4. While the full sky analysis—correctly—detects no significant

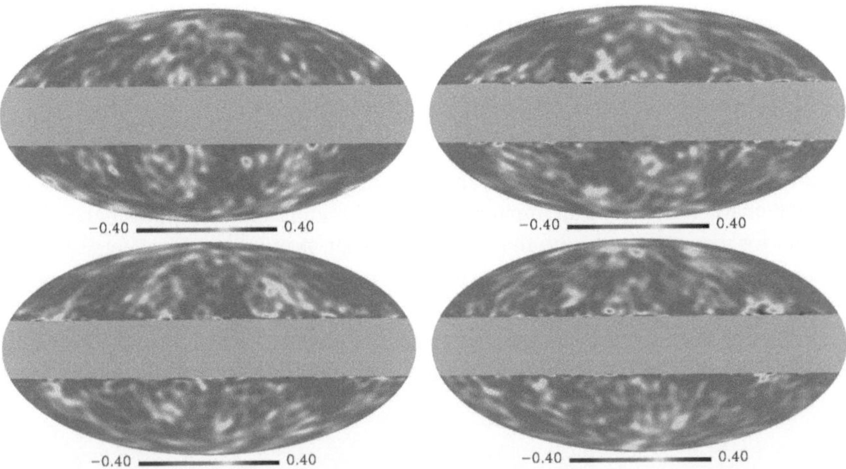

Fig. 2.4 A set of second order surrogates for the WMAP seven year ILC map (*upper left*) for a constant latitude cut, that removes $|b| < 20°$ of the Galactic plane. Here, the multipole limit is $\ell_{max} = 40$ (also for the original ILC map) and the shuffling range is chosen as $[\ell_1, \ell_2] = [2, 40]$. This represents the special case, where a shuffling outside the range is unnecessary, and therefore no first order surrogate exists

Fig. 2.5 The σ-normalised deviations between a simulated CMB map and its surrogate data sets. The colour of each pixel illustrates the mean deviation for the hemisphere around that pixel

deviations between the simulation and its surrogates, a clear shift to negative values, and therefore phase correlations, are identified for the cut sky cases. This shift is getting larger for increasing cuts, which clearly points towards a systematic effect.

A convincing analysis should therefore be able to remove these systematic effects. In Chap. 7, we will return to this problem and present an appropriate solution, thus enabling investigations by means of surrogates on an incomplete sky.

2.2 Measures for Non-Gaussianity

2.2.1 Overview Over Currently Used Measures

As soon as simulations or surrogates are created, one needs a measure being sensitive
to some characteristics of the input maps and providing output values that can then
be used for a comparison with the original data set. As we stated in Sect. 2.1.1, any
kind of possible deviation from as well as consistency with Gaussianity marks an
interesting result. Thus, one could think of a large amount of reasonable measures
that could be used for the comparison. In fact, a plethora of different measures has
been applied in CMB analysis until today. In general, these can be separated into
global and *local* measures.

Global Measures

Measures of global type are related to the characteristics of the map as a whole. One
of the currently most used measures is the *angular bispectrum* (e.g. [2, 9, 10]), which
is the harmonic transform of the three-point correlation function. Three different con-
figurations of the bispectrum are favoured. These depend on the shape of the triangle
describing the three-point correlation function, and are termed "local" (referring to
a "squeezed" triangle with two sides much larger than the third, [38]), "equilateral"
[39] and "orthogonal" [40]. The result for each configuration can be expressed as
one single value, the so-called *non-linear coupling parameter* f_{NL}, which describes
the amount of non-Gaussianity of the primordial gravitational potential: $f_{NL} = 0$
would refer to the Gaussian case, while any larger or smaller value points towards
deviations from Gaussianity. Both the parameter f_{NL} as well as the bispectrum can
be used in combination with other techniques, e.g. the bispectrum with the help of the
needlet coefficients (see also below), which is then referred to as *needlet bispectrum*
[41–43]. Another global measure is the *power spectrum*, which we defined already
on p. XXX, or the corresponding 2-point correlation function in real space. These
measures were applied in [44–50]. Although the power spectrum is not a measure
for non-Gaussianity (since it only analyses the Gaussian part of the temperature
anisotropies, see Sect. 1.3.3), it is listed here due to the important results concerning
asymmetries in the CMB sky: The power spectrum can be estimated using parts of
the sky only, hence modifying it to a measure of local type, which enabled the first
detection of power asymmetries in [46] (cf. Sect. 1.4). The next example for global
measures are *Minkowski functionals*. These are three related measures, which can be
interpreted as area, perimeter, and Euler parameter, that focus on geometrical structures in the data [9, 51]. In doing so, the map is grouped into active and inactive pixels
that are defined as pixels with lower/higher values than some given threshold. Then,
the structural behaviour of these two kinds of pixels is observed for different threshold values. In Chap. 7, we will use the Minkowski functionals on cut sky surrogates
parallel to an analysis by means of the scaling indices, that are described in more

detail in the following Sect. 2.2.2. Similar to the Minkowski functionals, the *length of the sceleton* describes an analysis that examines the length of the zero-contour line of the map, which is defined by derivatives of the field in different directions [51]. Another two global measures are the *genus* analysis [52] and *multipole vectors* [53–59]. The former investigates the different quantities of hot and cold spots, while the latter forms a set of unit vectors, that can be used to describe and analyse a given multipole ℓ. Finally, *phase mapping techniques* are a useful tool to detect deviations of the map from a Gaussian random field [24–31]. The basic assumption for this kind of analysis—that is independent and identically distributed phases—is the same as we used above for the construction of surrogates.

Local Measures

In contrast to the global measures, local measures investigate the behaviour of the maps in a direction-dependent way. This offers the possibility to identify the position of anomalies, which can then for example be compared with a large-scale structure survey. In addition, for investigations by means of local measures, one can simply exclude heavily foreground-affected regions like the Galactic plane.

A very famous example for a local measure in CMB analysis are *wavelets*. A wavelet is a filter function, that is used to transform the underlying map into wavelet space, where the structural behaviour of the data becomes more pronounced. For CMB analysis, *directional spherical wavelets* [8, 60, 61], *steerable wavelets* [1], and *spherical mexican hat wavelets* [62–66] have been applied, in which the investigation in [62] lead to the first detection of the famous Cold Spot (cf. Sect. 1.4). A very recently developed form of wavelets are spherical *needlets*, which allow to focus on a specific set of multipoles [7, 41–43]. As already stated above, needlets can also be used to construct the needlet bispectrum.

An analysis by means of *local curvature* classifies the map points by their type of curvature, that is hills, saddles and lakes [67]. Their distribution on the sphere can then be analysed and compared to that of simulations or surrogates. Similar to the power spectrum estimation from above, some measures have in general a global behaviour, but can be used as a local one by focusing on smaller regions on the sphere. While this usage is an exception for the power spectrum, it is common for the so-called *large-angle non-Gaussianity indicators* [68–70] and the *Kolmogorov stochasticity parameter* [71–73]. The former is based on skewness and kurtosis of the temperature values inside large-angle patches of CMB maps, while the latter examines the largest difference between theoretical and empirical cumulative distribution function. The technique of considering small caps on the sphere to transform a global measure to a local one was also applied for analyses of the angular two-point correlation function in [6].

2.2.2 The Scaling Index Method

The measure for non-Gaussianity of the CMB which is applied throughout this work is the *scaling index method* (SIM) [17, 18]. This measure has the ability of revealing the topological behaviour of an input map by detecting different structures in the data, as for example cluster-like or sheet-like structures, as well as filaments or walls. While wavelets are more sensitive to structures, which offer intensity variations of significant magnitude with respect to the existing noise, scaling indices also detect structural features which possess variations within the noise level, but not significantly higher or lower intensity values [74].

Scaling indices have already been used for texture discrimination [75] and feature extraction [76, 77], time series analysis of stock exchanges [78] and active galactic nuclei [79, 80], as well as structure analysis of bone images [81] and other different medical data, like biological specimens, skin cancer, computed tomographic images, and beat-to-beat sequences from electrocardiograms [16]. Investigations concerning the Gaussianity of the CMB by applying the SIM to simulated CMB maps and the WMAP three-year data were performed in [74] and [18], respectively, where the method turned out to be of great usefulness.

The basic ideas of the SIM stem from the calculation of dimensions of strange attractors in nonlinear time series analysis. If an attractor has a non-integer dimension, it is termed *strange* [82]. These attractors play an important role in the field of dynamical systems, since systems exhibiting chaotic behaviour often possess a strange attractor in phase space [82–85]. The dimension hereby provides information about the topological characteristics of the attractor [86–88].

The basis for the calculation of the dimension of attractors from a time series is to perform a transformation of the time series into a point distribution in d-dimensional Euclidian space [89]. This transformation and the d-dimensional space are also denoted *embedding* and *embedding space*, respectively. The most common example for such an embedding are the so-called *delay-coordinates* [90]. These are constructed from a time series x_i, $i = 1, ..., N$, of a single observed quantity from some experiment. The information of d data points can be combined to a vectors $\vec{p}_i$ in d-dimensional Euclidian space in the following way:

$$\vec{p}_i = (x_i, x_{i+\tau}, ..., x_{i+d\tau}) \ , \ i \in \{1, ..., N - d\tau\}$$

Here, the time interval τ which specifies the distance between the data points is termed delay time or lag. The resulting point set provides the analyst a completely new access for investigations of the data set. It was proven in [90], that the transformation to delay-coordinate maps is a diffeomorphism, that is a smooth invertible isomorph function with a smooth inverse that maps one differentiable manifold to another. Therefore, all the information of the time series is preserved. This result was extended to fractal sets in [89]. We use an approach analogously to this concept to enable the usage of the SIM on CMB data below.

After transforming the original data by means of such an embedding, and therefore obtaining a point set P with points $\vec{p}_i$, $i = 1, ..., N_p$ in Euclidian space, one can estimate the local scaling properties of this point set. In [83], this is done by counting the number of system states around one point $\vec{p}_i$ by means of the Heaviside function $H(x)$:

$$N(\delta, \vec{p}_i) = \sum_{j=1}^{N_p} H(\delta - \|\vec{p}_i - \vec{p}_j\|) , \tag{2.7}$$

where the Heaviside function is defined as $H(x) = 1$ for $x \geq 0$ and $H(x) = 0$ else. The parameter δ is used to set a boundary: If the distance $\|\vec{p}_i - \vec{p}_j\|$ is larger than δ, the resulting $H(x)$ becomes zero. The basic idea behind the setup of Eq. (2.7) is the following fact: For small r and a large amount of points N_p, the measure behaves as a power of r, with an exponent ν [16, 83]:

$$\lim_{N_p \to \infty} \frac{1}{N_p^2} N(\delta, \vec{p}_i) \propto \delta^\nu \tag{2.8}$$

The exponent ν is again closely related to the dimensionality of the strange attractor [83, 91]. Therefore, by calculating ν, one can obtain information about the topological characteristics of the attractor. This statement is also the crucial point for the scaling index approach, as we will see below. However, due to the discontinuity of the Heaviside function, the derivate of $H(x)$, and therefore also the exponent ν, cannot be evaluated analytically. One can only approximate ν by averaging over a chosen range $[\delta_1, \delta_2]$:

$$\nu \approx \frac{\log N(\delta_2, \vec{p}_i) - \log N(\delta_1, \vec{p}_i)}{\log \delta_2 - \log \delta_1} \tag{2.9}$$

The method explained above is not the only possible approach. Similar studies were considered e.g. in [86], where a one-dimensional return map was constructed from the embedding space. From this return map, one can evaluate the characteristic exponent of the attractor. In [92], the spectrum of singularities of scaling functions is computed, in order to describe the complex scaling of the attractor.

One can now modify these ideas to apply the scaling index approach to the CMB. Here, the fluctuations of the temperature map are characterised by the values of the pixelised sky on a sphere S. Thus, the analogue of an embedding for a CMB analysis, is a transformation of the combined temperature information and the two-dimensional spatial information on the sphere into a three-dimensional point set, which includes all the information of the original map as spatial information only. Here, the pixels (θ_i, ϕ_i), $i = 1, ..., N_{pix}$, of S, where N_{pix} denotes the number of pixels and (θ_i, ϕ_i) latitude and longitude of the pixel i on the sphere, are converted to a point distribution in a three-dimensional space in the following way: Each temperature value $T(\theta_i, \phi_i)$ is assigned to one point $\vec{p}_i$, which is located in the radial direction through its pixel's centre (θ_i, ϕ_i), that is a straight line perpendicular to the surface of the sphere. Thus, the three-dimensional position vector of the new point $\vec{p}_i$ reads as

$$x_i = (R + dR)\cos(\phi_i)\sin(\theta_i) \tag{2.10}$$

$$y_i = (R + dR)\sin(\phi_i)\sin(\theta_i) \tag{2.11}$$

$$z_i = (R + dR)\sin(\theta_i) \tag{2.12}$$

with

$$dR = a\left(\frac{T(\theta_i, \phi_i) - \langle T \rangle}{\sigma_T}\right), \tag{2.13}$$

where R denotes the radius of the sphere and a describes an adjustment parameter. In addition, $\langle T \rangle$ and σ_T characterise the mean and the standard deviation of the temperature fluctuations, respectively. The normalisation is performed to obtain for dR zero mean and a standard deviation of a. A transformed CMB map appearing as a three-dimensional point distribution is illustrated in Fig. 2.6. Here, two different values for a were used in the embedding process.

In general, the SIM is—like ν in Eq. (2.8)—a mapping that calculates for every point $\vec{p}_i$ of the point set P a single value, which depends on the spatial position of $\vec{p}_i$ in the group of the other points. P is three-dimensional for this chosen embedding of CMB data. For every point $\vec{p}_i$, we define the *local weighted cumulative point distribution* as

$$\rho(\vec{p}_i, r) = \sum_{j=1}^{N_{pix}} s_r(d(\vec{p}_i, \vec{p}_j))$$

with r describing the scaling range (similar to δ in Eq. (2.7)), while $s_r(\bullet)$ and $d(\bullet)$ denote a differentiable shaping function and a distance measure, respectively. To obtain the *scaling index* $\alpha(\vec{p}_i, r)$, we assume the following scaling law, which is similar to Eq. (2.8):

$$\rho(\vec{p}_i, r) \propto r^{\alpha(\vec{p}_i, r)} \tag{2.14}$$

One important difference to the above concept is the request for a differentiable shaping function $s_r(\bullet)$, which leads also to a differentiable cumulative point distribution $\rho(\vec{p}_i, r)$. Therefore, in contrast to Eq. (2.8) above, the scaling law (2.14) becomes analytically solvable. The scaling index, corresponding to the exponent ν in Eq. (2.8), can therefore be computed as the logarithmic derivative of $\rho(\vec{p}_i, r)$. If we choose e.g. Gaussian shaping functions

$$s_r(x) = e^{-\left(\frac{x}{r}\right)^n},$$

the scaling index reads as

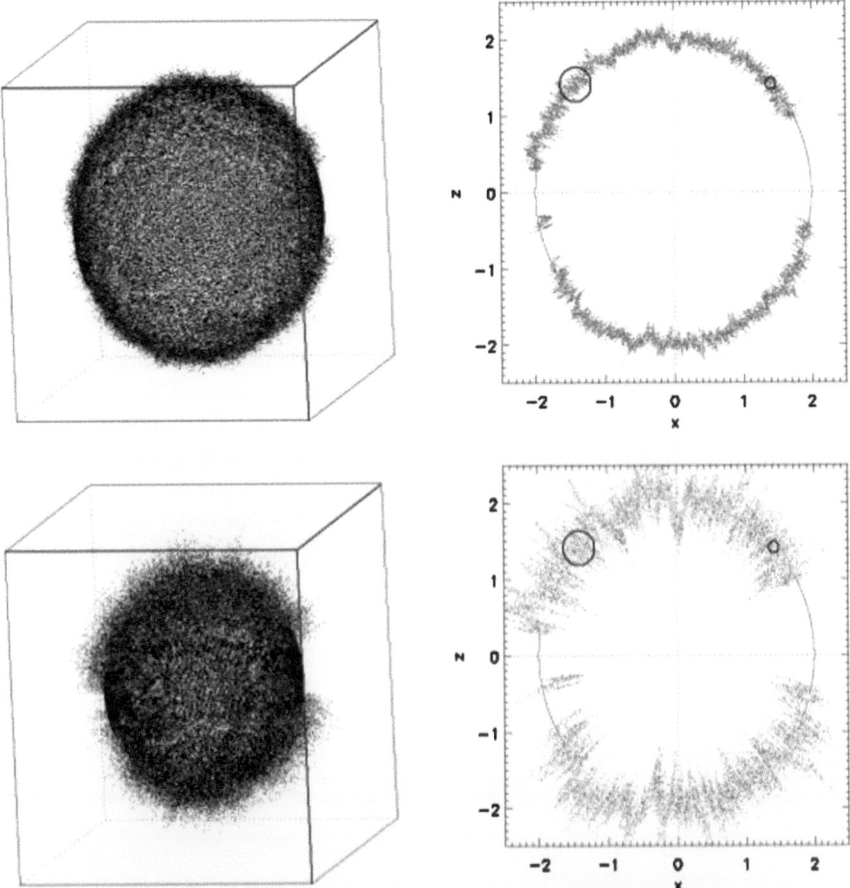

Fig. 2.6 WMAP 3-year data after application of the transformation into a three-dimensional point distribution. On the left side the full set of points is presented, while the right side shows an x, z-projection of only the points with $|y| < 0.05$. In other words, the plots show the "roughness" of the last scattering surface. Two different values for a were used, namely $a = 0.075$ (*above*) and $a = 0.225$ (*below*). The black circles represent the scaling ranges $r = 0.075$ and $r = 0.225$. Figure taken from [74]

$$\alpha(\vec{p}_i, r) = \frac{\partial \log \rho(\vec{p}_i, r)}{\partial \log r} = \frac{\sum_{j=1}^{N_{pix}} n\left(\frac{d(\vec{p}_i, \vec{p}_j)}{r}\right) e^{-\left(\frac{d(\vec{p}_i, \vec{p}_j)}{r}\right)^n}}{\sum_{j=1}^{N_{pix}} e^{-\left(\frac{d(\vec{p}_i, \vec{p}_j)}{r}\right)^n}} \, .$$

In general, one can freely choose $s_r(\bullet)$ and $d(\bullet)$, apart from the requirement that $s_r(\bullet)$ has to be differentiable. For the analysis in this work, we make use of a set of quadratic Gaussian shaping functions and the isotropic Euclidian norm as distance measure:

$$s_r(x) = e^{-\left(\frac{x}{r}\right)^2}$$

$$d(\vec{p}_i, \vec{p}_j) = \|\vec{p}_i - \vec{p}_j\|_2$$

Taking this into account, and using in addition the abbreviation $d_{ij} := \|\vec{p}_i - \vec{p}_j\|_2$, we obtain the final formula of the scaling indices:

$$\alpha(\vec{p}_i, r) = \frac{\sum_{j=1}^{N_{pix}} 2\left(\frac{d_{ij}}{r}\right)e^{-\left(\frac{d_{ij}}{r}\right)^2}}{\sum_{j=1}^{N_{pix}} e^{-\left(\frac{d_{ij}}{r}\right)^2}} \tag{2.15}$$

In the resulting map $\alpha(\vec{p}_i, r)$, $i = 1, ..., N_{pix}$, the structural behaviour of the underlying point set P becomes apparent, and different types of structure can be detected very easily. The values of α are related to structural characteristics in the following way: A point- or cluster-like structure leads to scaling indices $\alpha \approx 0$, filaments to $\alpha \approx 1$ and sheet-like structures to $\alpha \approx 2$. A uniform distribution of points would result in $\alpha \approx 3$. In between, curvy lines and curvy sheets produce $1 \leq \alpha \leq 2$ and $2 \leq \alpha \leq 3$, respectively. Underdense regions in the vicinity of point-like structures, filaments or walls feature $\alpha > 3$. An example of a simulated CMB map and its scaling index response is shown in Fig. 2.7.

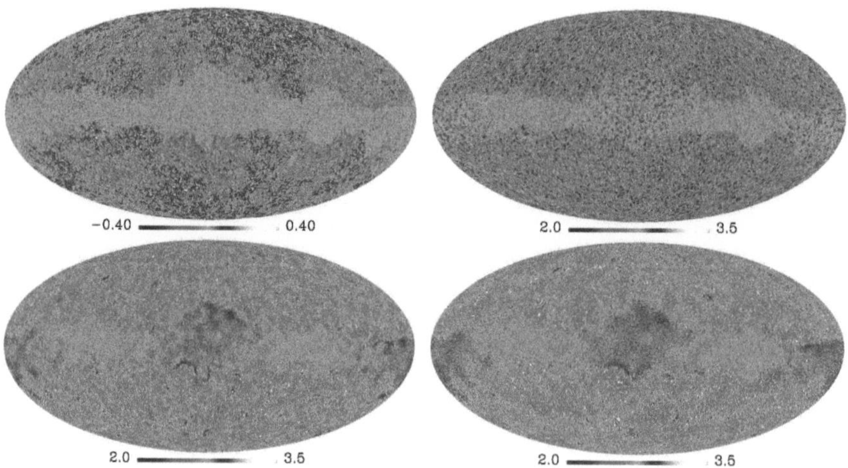

Fig. 2.7 A simulated CMB map, in which the central regions were masked out and filled with noise whose variance corresponds to the noise characteristics of the WMAP satellite (*upper left*), and the scaling index responses $\alpha(\vec{p}_i, r)$ for three different scaling ranges: $r = 0.05$ (*upper right*), $r = 0.15$ (*lower left*) and $r = 0.25$ (*lower right*). Different values of $\alpha(\vec{p}_i, r)$ correspond to different types of structure in the underlying map. Small scaling ranges examine the behaviour of the small structures, while the characteristics of the larger structure is displayed by the higher scaling ranges. Note the different structures inside and outside the masked region of the simulated map, and also the different structures in the mask itself due to the noise characteristics. Both is clearly identified by the scaling indices

From Eq. (2.15), one can see that the scaling range parameter r can be chosen arbitrarily. This parameter weights the distances between our point of interest $\vec{p}_i$ and the remaining points $\vec{p}_j$ (cf. definition of $s_r(x)$). Therefore, we can make use of smaller or larger values for r to examine the different behaviour of the small-scale or large-scale structural configuration in the underlying map. For the analysis in the following Chapters, we mostly make use of the ten scaling range parameters $r_k = 0.025, 0.05, ..., 0.25, k = 1, 2, ...10$. Table 4.1 on p. XXX illustrates how the positions of the resulting 90 % and 10 % weighting of the quadratic Gaussian shaping function $s_r(x)$ correspond to the angular scale ℓ in Fourier space. In Fig. 2.7, three different values of r were applied to the simulated CMB map.

In addition, both R and a from the Eqs. (2.10) and (2.13) should be chosen in a proper way to ensure a high sensitivity of the SIM with respect to the temperature fluctuations at a certain spatial scale. For CMB analysis, it turned out that this requirement is provided using $R = 2$ for the radius of the sphere and setting the parameter a, which describes the standard deviation of the normalised temperature values, to the value of the scaling range parameter r [74]. Thus, in this case the distance $1r$ corresponds to 1σ of the temperature distribution.

When we apply the scaling index method to CMB data sets, there are different methods of how to compare the results with those from simulations or surrogate maps. On the one hand, one can carry out a global analysis by calculating statistics like the mean or the standard deviation,

$$\langle \alpha_r \rangle = \frac{1}{N_{pix}} \sum_{i=1}^{N_{pix}} \alpha(\vec{p}_i, r)$$

$$\sigma_{\alpha_r} = \left(\frac{1}{N_{pix} - 1} \sum_{i=1}^{N_{pix}} \left[\alpha(\vec{p}_i, r) - \langle \alpha_r \rangle \right]^2 \right)^{1/2} ,$$

of the scaling index response for the complete set of pixels. On the other hand, it is also possible to perform a local analysis by focusing on a particular area, as for example a hemisphere that is located in some chosen direction on the sphere. These methods will be repeatedly applied throughout Chaps. 4–7.

References

1. P. Vielva, E. Martínes-González, M. Cruz, R.B. Barreiro, M. Tucci, Monthly Notices of the Royal Astronomical Society, **410**(1), 33 (2011)
2. E. Komatsu, K.M. Smith, J. Dunkley, C.L. Bennett, B. Gold, G. Hinshaw, N. Jarosik, D. Larson, M.R. Nolta, L. Page, D.N. Spergel, M. Halpern, R.S. Hill, A. Kogut, M. Limon, S.S. Meyer, N. Odegard, G.S. Tucker, J.L. Weiland, E. Wollack, E.L. Wright, ApJS **192**, 18 (2011)
3. D. Munshi, P. Coles, A. Cooray, A. Heavens, J. Smidt, MNRAS **410**, 1295 (2011)
4. E.F. Bunn, in *Proceedings of Rencontres de Moriond* (2010)

5. C.L. Bennett, R.S. Hill, G. Hinshaw, D. Larson, K.M. Smith, J. Dunkley, B. Gold, M. Halpern, N. Jarosik, A. Kogut, E. Komatsu, M. Limon, S.S. Meyer, M.R. Nolta, N. Odegard, L. Page, D.N. Spergel, G.S. Tucker, J.L. Weiland, E. Wollack, E.L. Wright, ApJS **192**, 17 (2011)
6. A. Bernui, Phys. Rev. D **78**, 063531 (2008)
7. D. Pietrobon, A. Amblard, A. Balbi, P. Cabella, A. Cooray, D. Marinucci, Phys. Rev. D **78**, 103504 (2008)
8. J.D. McEwen, M.P. Hobson, A.N. Lasenby, D.J. Mortlock, MNRAS **388**, 659 (2008)
9. E. Komatsu, A. Kogut, M.R. Nolta, C.L. Bennett, M. Halpern, G. Hinshaw, N. Jarosik, M. Limon, S.S. Meyer, L. Page, D.N. Spergel, G.S. Tucker, L. Verde, E. Wollack, E.L. Wright, ApJS **148**, 119 (2003)
10. K.M. Smith, L. Senatore, M. Zaldarriaga, JCAP **0909**, 006 (2009)
11. J. Theiler, S. Eubank, A. Longtin, B. Galdrikian, J.D. Farmer, Physica D **58**, 77 (1992)
12. T. Schreiber, A. Schmitz, Physica D **142**, 346 (2000)
13. T. Schreiber, A. Schmitz, Phys. Rev. Lett. **77**, 635 (1996)
14. T. Schreiber, A. Schmitz, Phys. Rev. Lett. **80**, 2105 (1998)
15. M.P. Pompilio, F.R. Bouchet, G. Murante, A. Provenzale, ApJ **449**, 1 (1995)
16. A. Bunde, J. Kropp, H.-J. Schellnhuber, *The Science of Disasters: Climate Disruptions, Heart Attacks, and Market Crashes* (Springer, Berlin, 2002)
17. C. Räth, W. Bunk, M.B. Huber, G.E. Morfill, J. Retzla, P. Schuecker, MNRAS, **337**, 413 (2002)
18. C. Räth, P. Schuecker, MNRAS, **344**, 115 (2003)
19. E. Komatsu, N. Afshordi, N. Bartolo, D. Baumann, J.R. Bond, E.I. Buchbinder, C.T. Byrnes, X. Chen, D.J.H. Chung, A. Cooray, P. Creminelli, N. Dalal, O. Dore, R. Easther, A.V. Frolov, K.M. Górski, M.G. Jackson, J. Khoury, W.H. Kinney, L. Kofman, K. Koyama, L. Leblond, J.-L. Lehners, J.E. Lidsey, M. Liguori, E.A. Lim, A. Linde, D.H. Lyth, J. Maldacena, S. Matarrese, L. McAllister, P. McDonald, S. Mukohyama, B. Ovrut, H.V. Peiris, C. Raeth, A. Riotto, Y. Rodriguez, M. Sasaki, R. Scoccimarro, D. Seery, E. Sefusatti, U. Seljak, L. Senatore, S. Shandera, E.P.S. Shellard, E. Silverstein, A. Slosar, K.M. Smith, A.A. Starobinsky, P.J. Steinhardt, F. Takahashi, M. Tegmark, A.J. Tolley, L. Verde, B.D. Wandelt, D. Wands, S. Weinberg, M. Wyman, A.P.S. Yadav, M. Zaldarriaga, Astro2010: The astronomy and astrophysics decadal survey. Sci. White Pap. **158**(2009)
20. L.-Y. Chiang, P. Coles, P. Naselsky, MNRAS **337**, 488 (2002)
21. P.D. Naselsky, O.V. Verkhodanov, Int. J. Mod. Phys. D **17**, 179 (2008)
22. P. Naselsky, D. Novikov, J. Silk, ApJ **565**, 655 (2002)
23. P.D. Naselsky, I.D. Novikov, L.-Y. Chiang, ApJ **642**, 617 (2006)
24. L.-Y. Chiang, P.D. Naselsky, O.V. Verkhodanov, M.J. Way, ApJ Lett. **590**, L65 (2003)
25. P.D. Naselsky, A.G. Doroshkevich, O.V. Verkhodanov, ApJ Lett. **599**, L53 (2003)
26. P.D. Naselsky, L.-Y. Chiang, P. Olesen, O.V. Verkhodanov, ApJ **615**, 45 (2004)
27. P.D. Naselsky, L.-Y. Chiang, P. Olesen, I. Novikov, Phys. Rev. D **72**, 063512 (2005)
28. P. Coles, P. Dineen, J. Earl, D. Wright, MNRAS **350**, 989 (2004)
29. L.-Y. Chiang, P.D. Naselsky, Int. J. Mod. Phys. D, **15**, 1283 (2006)
30. L.-Y. Chiang, P.D. Naselsky, P. Coles, ApJ **664**, 8 (2007)
31. L.-Y. Chiang, P.D. Naselsky, MNRAS Lett. **380**, L71 (2007)
32. B. Gold, N. Odegard, J.L. Weiland, R.S. Hill, A. Kogut, C.L. Bennett, G. Hinshaw, X. Chen, J. Dunkley, M. Halpern, N. Jarosik, E. Komatsu, D. Larson, M. Limon, S.S. Meyer, M.R. Nolta, L. Page, K.M. Smith, D.N. Spergel, G.S. Tucker, E. Wollack, E.L. Wright, ApJS **192**, 15 (2011)
33. J. Delabrouille, J.-F. Cardoso, M. Le Jeune, M. Betoule, G. Fay, F. Guilloux, A&A **493**, 835 (2009)
34. K.M. Górski, ApJ Lett. **430**, L85 (1994)
35. K.M. Górski, ApJ Lett. **430**, L89 (1994)
36. D.J. Mortlock, A.D. Challinor, M.P. Hobson, MNRAS **330**, 405 (2002)
37. W.H. Press, S.A. Teukolsky, W.T. Vetterling, B.P. Flannery, *Numerical Recipes: The Art of Scientific Computing*, 3rd edn. (Cambridge University Press, 2007)
38. D. Babich, P. Creminelli, M. Zaldarriaga, JCAP **0408**, 009 (2004)
39. P. Creminelli, A. Nicolis, L. Senatore, M. Tegmark, M. Zaldarriaga, JCAP **0605**, 004 (2006)

40. L. Senatore, K.M. Smith, M. Zaldarriaga, JCAP **1**, 28 (2010)
41. O. Rudjord, F.K. Hansen, X. Lan, M. Liguori, D. Marinucci, S. Matarrese, ApJ **701**, 369 (2009)
42. O. Rudjord, F.K. Hansen, X. Lan, M. Liguori, D. Marinucci, S. Matarrese, ApJ **708**, 1321 (2010)
43. P. Cabella, D. Pietrobon, M. Veneziani, A. Balbi, R. Crittenden, G. de Gasperis, C. Quercellini, N. Vittorio, MNRAS **405**, 961 (2010)
44. I.J. O'Dwyer, H.K. Eriksen, B.D. Wandelt, J.B. Jewell, D.L. Larson, K.M. Górski, A.J. Banday, S. Levin, P.B. Lilje, ApJ Lett. **617**, L99 (2004)
45. H.K. Eriksen, G. Huey, A.J. Banday, K.M. Górski, J.B. Jewell, I.J. O'Dwyer, B.D. Wandelt, ApJ **665**, 1L (2007)
46. H.K. Eriksen, F.K. Hansen, A.J. Banday, K.M. Górski, P.B. Lilje, ApJ **605**, 14 (2004)
47. F.K. Hansen, A.J. Banday, K.M. Górski, MNRAS **354**, 641 (2004)
48. F.K. Hansen, A.J. Banday, K.M. Górski, H.K. Eriksen, P.B. Lilije, ApJ **704**, 1448 (2009)
49. J. Hoftuft, H.K. Eriksen, A.J. Banday, K.M. Górski, F.K. Hansen, P.B. Lilje, ApJ **699**, 985 (2009)
50. F. Paci, A. Gruppuso, F. Finelli, P. Cabella, A. De Rosa, N. Mandolesi, P. Natoli, MNRAS **407**, 339 (2010)
51. H.K. Eriksen, D.I. Novikov, P.B. Lilje, A.J. Banday, K.M. Górski, ApJ **612**, 64 (2004)
52. C.-G. Park, MNRAS **349**, 313 (2004)
53. D.J. Schwarz, G.D. Starkman, D. Huterer, C.J. Copi, Phys. Rev. Lett. **93**, 221301 (2004)
54. C.J. Copi, D. Huterer, D.J. Schwarz, G.D. Starkman, MNRAS **367**, 79 (2006)
55. C.J. Copi, D. Huterer, D.J. Schwarz, G.D. Starkman, Phys. Rev. D **75**, 023507 (2007)
56. C.J. Copi, D. Huterer, D.J. Schwarz, G.D. Starkman, Adv. in Astr., 847541 (2010)
57. A. Gruppuso, C. Burigana, JCAP **0908**, 004 (2009)
58. A. Gruppuso, K.M. Górski, JCAP **03**, 019 (2010)
59. D. Sarkar, D. Huterer, C.J. Copi, G.D. Starkman, D.J. Schwarz, Astropart. Phys. **34**, 591 (2011)
60. J.D. McEwen, M.P. Hobson, A.N. Lasenby, D.J. Mortlock, MNRAS **359**, 1583 (2005)
61. J.D. McEwen, M.P. Hobson, A.N. Lasenby, D.J. Mortlock, MNRAS Lett. **371**, L50 (2006)
62. P. Vielva, E. Martínes-González, R.B. Barreiro, J.L. Sanz, L. Cayón, ApJ **609**, 22 (2004)
63. P. Mukherjee, Y. Wang, ApJ **613**, 51 (2004)
64. M. Cruz, E. Martínes-González, P. Vielva, L. Cayón, MNRAS **356**, 29 (2005)
65. M. Cruz, M. Tucci, E. Martínes-González, P. Vielva, MNRAS **369**, 57 (2006)
66. M. Cruz, L. Cayón, E. Martínes-González, P. Vielva, J. Jin, ApJ **655**, 11 (2007)
67. F.K. Hansen, P. Cabella, D. Marinucci, N. Vittorio, ApJ Lett. **607**, L67 (2004)
68. A. Bernui, M.J. Reboucas, Int. J. Mod. Phys. D **19**, 1411 (2010)
69. A. Bernui, M.J. Reboucas, A.F.F. Teixeira, Int. J. Mod. Phys. D **19**, 1405 (2010)
70. A. Bernui, M.J. Reboucas, Phys. Rev. D **81**, 063533 (2010)
71. V.G. Gurzadyan, A.A. Kocharyan, A&A Lett. **492**, L33 (2008)
72. V.G. Gurzadyan, A.E. Allahverdyan, T. Ghahramanyan, A.L. Kashin, H.G. Khachatryan, A.A. Kocharyan, H. Kuloghlian, S. Mirzoyan, E. Poghosian, G. Yegorian, A&A **497**, 343 (2009)
73. V.G. Gurzadyan, A.L. Kashin, H.G. Khachatryan, A.A. Kocharyan, E. Poghosian, D. Vetrugno, G. Yegorian, Europhys. Lett. **91**, 19001 (2010)
74. C. Räth, P. Schuecker, A.J. Banday, MNRAS **380**, 466 (2007)
75. C. Räth, G.E. Morfill, JOSA A **14**, 3208 (1997)
76. F. Jamitzky, R.W. Stark, W. Bunk, S. Thalhammer, C. Räth, T. Aschenbrenner, G.E. Morfill, W.M. Heckl, Ultramicroscopy **86**, 241 (2001)
77. C. Räth, R. Monetti, J. Bauer, I. Sidorenko, D. Müller, M. Matsuura, E.-M. Lochmüller, P. Zysset, F. Eckstein, New J. Phys. **10**, 125010 (2008)
78. R. Monetti, H. Boehm, D. Müller, E. Rummeny, T. Link, C. Räth, Proc. SPIE **5370**, 215 (2004)
79. M. Gliozzi, W. Brinkmann, C. Räth, I.E. Papadakis, H. Negoro, H. Scheingraber, A&A **391**, 875 (2002)
80. M. Gliozzi, I.E. Papadakis, C. Räth, A&A **449**, 969 (2006)
81. D. Müeller, T. Link, R. Monetti, J. Bauer, H. Boehm, V. Seifert-Klauss, E. Rummeny, G. Morfill, C. Räth, Osteo. Int. **17**, 1483 (2006)

82. D. Ruelle, E. Takens, Commun. Math. Phys. **50**, 69 (1976)
83. P. Grassberger, I. Procaccia, PRL **50**, 346 (1983)
84. E.N. Lorenz, J. Atmos, Science **20**, 130 (1963)
85. E. Ott, Rev. Mod. Phys. **53**, 655 (1981)
86. N.H. Packard, J.P. Crutchfield, J.D. Farmer, R.S. Shaw, PRL **45**, 712 (1980)
87. J.C. Kaplan, J.A. Yorke, *Functional Differential Equations and Approximations of Fixed Points* (Springer, Berlin, 1979)
88. H. Mori, Prog. Theor. Phys. **63**, 1044 (1980)
89. T. Sauer, J.A. Yorke, M. Casdagli, J. Stat. Phys. **65**, 579 (1991)
90. F. Takens, in *Dynamical Systems and Bifurcations*, ed. by D.A. Rand, L.-S. Young. Detecting strange attractors in turbulence, Lecture Notes in Mathematics, vol 898, (Springer, Berlin, 1981), p. 366
91. P. Grassberger, R. Badii, A. Politi, J. Stat. Phys. **51**, 135 (1988)
92. T. Halsey, M.H. Jensen, L.P. Kadanoff, I. Procaccia, B.I. Shraiman, Phys. Rev. A **33**, 1141 (1986)

Chapter 3
Observations of the CMB
with the WMAP Satellite

3.1 Framework of the Observation

The academic history of the CMB is a rather young story: The existence of a relic radiation was first proposed by George Gamov in 1946 [1], and the evaluation of its temperature was done by Ralph Alpher and Robert Herman in 1950 [2]. The discovery of the radiation by Arno Penzias and Robert Wilson in 1965 represents a milestone in the history of Cosmology [3]. The first measurement with a full-sky coverage was realised with the satellite *Cosmic Background Explorer* (COBE) that was launched in 1989. This space mission also detected—for the very first time—the small anisotropies of the CMB [4]. Although there are a lot of ground based projects, like e.g. the South Pole Telescope [5], the Saskatoon [6] and Python [7] teleskopes, the Tenerife Experiment [8], COSMOSOMAS [9], the Very Small Array [10] or the Boomerang balloon [11], the most commonly used data sets to date stem from the *Wilkinson Microwave Anisotropy Probe* (WMAP) [12], which measures the CMB with a very high accuracy. Some ground based observations like ACBAR [13] or the Cosmic Background Imager [14] feature a higher resolution at smaller scales. The PLANCK probe, launched in 2009, will soon succeed WMAP by providing even more precise full-sky measurements of the microwave background [15]. Figure 3.1 illustrates the different maps of the full-sky surveys up to today.

The WMAP satellite was launched in June 2001 and orbits the Sun-Earth Lagrange point L2 at a distance of around 1.5 million kilometers from Earth [16]. The probe contains 20 differential radiometers that are passively cooled to around 90 K by solar panels, which are always orientated towards the sun. The radiometers cover the five frequency bands 20–25 GHz, 28–36 GHz, 35–46 GHz, 53–69 GHz, and 82–106 GHz, which are denoted as K-, Ka-, Q-, V-, and W-band, respectively. Two radiometers are arranged in the former two bands K and Ka each, while both the Q- and V-band contain four radiometers. The remaining eight radiometers belong to the W-band. In turn, each two radiometers form one differencing assembly. Eventually, we obtain ten differencing assemblies which are identified as K, Ka, Q1, Q2, V1, V2, W1, W2, W3 and W4. The WMAP team provides the data as full-sky temperature maps per band (as shown in Fig. 3.1) and also per differencing assembly [17].

G. Rossmanith, *Non-linear Data Analysis on the Sphere*, Springer Theses, DOI: 10.1007/978-3-319-00309-2_3, © Springer International Publishing Switzerland 2013

Fig. 3.1 Four different full-sky maps of the CMB anisotropies: As it would have been seen by the observers Penzias and Wilson in 1965 (*upper left*), as it was measured by the COBE satellite in 1992 (*upper right*), and as it is measured today by the WMAP satellite for the low-frequency K- (*lower left*) and the high-frequency W-band (*lower right*), respectively. Both WMAP figures are based on the current seven-year data. The colour-coded temperature values range from $-100\,\mu$K to $100\,\mu$K for the COBE and -200μK to $200\,\mu$K for the WMAP data. The map of Penzias and Wilson is for demonstration purposes only, and hence possesses no colour-coding. All figures were taken from [17] with acknowledgements to the WMAP team

The WMAP observations are accomplished with a resolution of <13.8 arcmin FWHM [18]. The resulting data are provided as pixelised sky maps. The ordering follows the HEALPix scheme [19, 20]. Thereby, the sky is divided into twelve squares covering the same size: Four attached to the Galactic north pole, four attached to the according south pole, and four arranged around the equatorial line. Each square is then again filled with $N_{side} \times N_{side}$ equal-sized pixels, with N_{side} being an arbitrary power of 2. Hence, the total number of map pixels N_{pix} is obtained as $N_{pix} = 12N_{side}^2$. The standard maps of the WMAP team feature a resolution of $N_{side} = 512$, which corresponds to a pixel number of $N_{pix} = 3,145,728$. However, a couple of analyses are carried out by using only a reduced resolution of $N_{side} = 256$ or even less, most of the times for computational reasons or to easily remove noise effects that appear on very small scales (e.g. [21–23]).

The earliest data release of the WMAP team was made publicly available in 2003 and contained the observations of the first year. Since then, there were another three releases, including the current seven-year data set published in 2010.

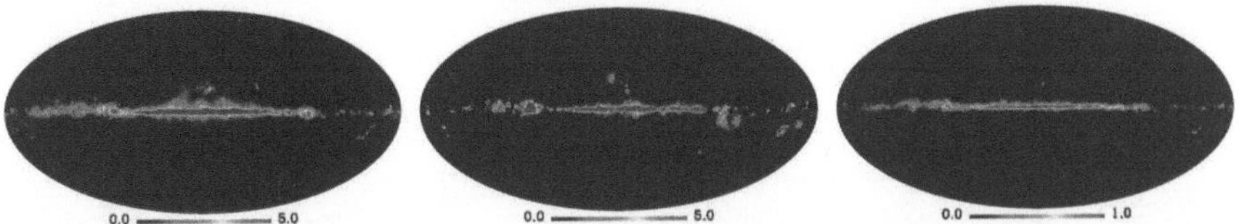

Fig. 3.2 The three primary foreground effects as measured by WMAP, from *left* to *right*: Synchrotron emission, free-free radiation, and dust emission. Every plot represents the effect on the band that it affects the most, see also Fig. 3.3. Therefore, the former two represent the foreground influences on the K-band, while the latter one the effect on the W-band. Note that the scale (in mK) for the dust emission is smaller

3.2 Foreground and Systematic Effects

3.2.1 Origin and Characteristics

The WMAP probe measures the microwave background radiation at a very high accuracy. These measurements get distorted by *foreground effects*. There are basically three primary mechanisms [24]: Synchrotron emission, free-free radiation (also known as thermal Bremsstrahlung) and dust emission. The intensities of each of the three foreground effects are illustrated in Fig. 3.2. The WMAP satellite observes at frequencies that are very close to the interval, where the CMB anisotropies are the highest in comparison with the fluctuations due to distortions [24]. Nevertheless, the effects of the foregrounds are still significant. The strongest influences of this type are caused by diffuse emission due to the Galactic plane.

When relativistic electrons interact with the Galactic magnetic field, *Synchrotron emission* is produced [24, 25]. The magnetic field forces the electrons to spiral and is thus changing their velocity, which causes the emission of radiation. Typical values of such a magnetic field are a few micro-Gauss. Synchrotron emission dominates in the low frequency band K, which is therefore the best band to detect it. The intensity of this emission decreases when going to higher frequency values. In the V and W bands, synchrotron emission is already very weak. A similar behaviour occurs for the *free-free radiation*. This radiation appears due to less energetic electrons, which scatter with ions or with each other. The electrons are decelerated, thus radiation is produced. I can be approximated with the use of $H\alpha$ emission [26]. The free-free radiation is never the dominant source in the measurements. While the two yet mentioned effects appear mainly at the lower frequencies, the contrary holds for the *dust grains*, which influences primarily the high frequency band W. Dust can be heated by ambient radiation, which is then re-emitted as radiation. In addition, dust could emit radiation due to rotational modes or excitations of their vibrational modes [27–29]. Due to these excitations, dust grains can also lead to a modified blackbody spectrum [30]. The effects of dust can be estimated by extrapolation based on dust models at higher frequencies. The frequency dependence of all these different foregrounds is shown in Fig. 3.3.

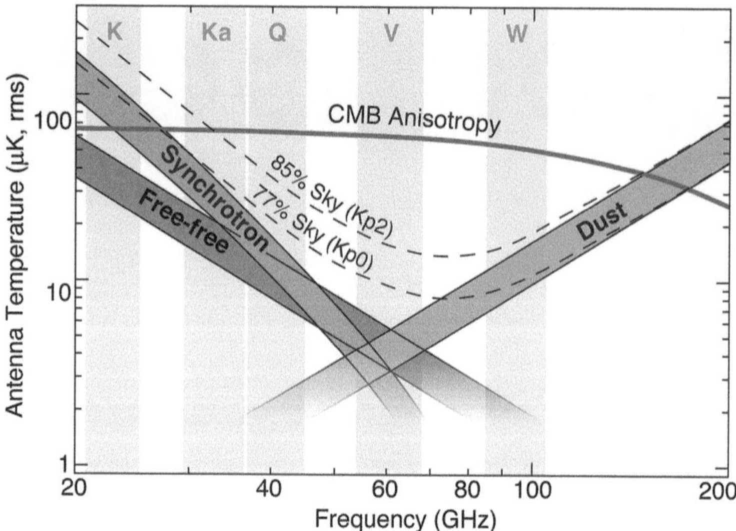

Fig. 3.3 The influence of the three primary foreground effects onto the different frequency bands of WMAP. Figure taken from [17] with acknowledgements to the WMAP team

In addition to foreground effects, *artificial effects* can distort the measurements, e.g. caused by possible systematic errors in the rather complex map-making process. While this is of course tried to be ruled out, and therefore subject to a multiplicity of analyses of the WMAP team [31], there are indeed doubts about the correctness of some details of the systematics, as for example about the proper way of removing the Doppler effect induced by the joint motion of the solar system and the space-craft [32–34]. These claims are again in parts questioned, but to some extend also confirmed in [35, 36].

3.2.2 *Methods of Foreground Reduction*

There are different techniques on how to handle and overcome the problem of fore-ground effects in the CMB signal. A very direct approach is to create templates for the three different foreground effects from above. This is done for the WMAP data by means of the foreground template model of [37, 38]. After establishing the fore-ground templates, one can subtract them from the original measurements. Thereby, the noise effects remain in the map. These foreground-cleaned maps are provided by the WMAP team [17]. Since the K-band and most of the time also the Ka-band are used in the foreground reduction process, the cleaned maps are mostly only available for the remaining Q-, V- and W-bands. Two of them are shown in the upper row of Fig. 3.4.

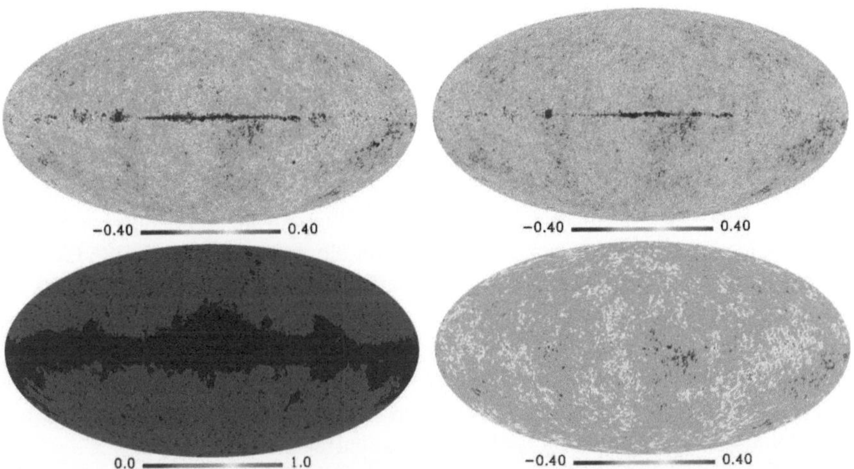

Fig. 3.4 Three different techniques of handling distortions in the CMB: Foreground reduction by means of template models, here illustrated by the resulting cleaned Q- and W-band maps (*upper row*), the KQ75 mask (*lower left*), which attempts to cut out all foreground-affected regions, and the ILC map (*lower right*), which is a weighted linear combination of the different frequency bands. All maps are based on the current seven-year data. Except for the lower left plot, the scale is in mK

However, the behaviour of contortions inside and close to the Galactic plane is still not completely understood. On the other hand, the major part of the measured sky can be taken as not affected by foregrounds [24]. Thus, the idea of just cutting out the highly distorted regions around the Galaxy is sometimes a much more helpful possibility of dealing with the foreground problem. For this reason, the WMAP team provides two masks: The KQ85 and the KQ75 masks, that—in their current version corresponding to the seven year data release—cut out 78.3 % and 70.6 % of the whole sky, respectively. The latter KQ75 mask is shown in the lower left plot of Fig. 3.4. The marginal difference between the two masks is simply due to different settings on how conservative the foregrounds should be treated. Although the foreground contribution is most intense in the Galactic plane, point sources affect the observation as well. In the current seven year data release, WMAP measured 471 point sources [24]. These are included into the masks as a circular cut-out of 1° for each source, except for the Centaurus A galaxy, which is cut out by a 3° circle. For analyses of the CMB that need a signal as clean as possible but do not suffer that much from a lower sky coverage, the use of masks is more advantageous than foreground reduction. In particular, this situation often occurs for investigations that search for non-Gaussianities with the help of a local measure, like the ones listed in Sect. 2.2.1 above. These analyses can simply leave out the masked parts of the sky. However, this approach can still lead to distorting effects at the border of the mask, which can again be compensated with techniques like a mask-filling method, as we will see in more detail in the following Chap. 4.

Other analyses prefer or even need a full-sky coverage, though. Especially investigations dealing with the spherical harmonics often require a complete sphere: As it can be seen in Fig. 3.4, the reduction process by means of foreground templates not always leads to satisfying results, since the Galactic plane is still apparent. For this reason, there is another technique of obtaining a more realistic CMB signal with low foreground influences, namely the Internal Linear Combination (ILC) method [24]. The basic idea is to combine the measurements of all frequency bands. The sky is divided into twelve fractions, whereupon eleven are located in the Galactic plane and only one in the minor foreground-affected remains. Different weights for the different bands are determined so as to minimise the variance of the temperature fluctuations at one degree resolution. For every fraction of the sky, a separate set of weights is estimated. For the final ILC map, the boundaries between these fractions are smoothed with a $1.5°$ kernel. One obtains a map with low foreground influence and a reliable CMB estimation for large-scale analyses.

The ILC map provided by the WMAP team is not the only map based on the concept of combining the different bands. An example for a related map with even lower amount of foreground effects is the Needlet-based Internal Linear Combination (NILC) map of [39], which is based on the WMAP five-year data. For the NILC map, the contamination by noise and foregrounds is minimised by means of a (one-dimensional) Wiener filtering. The important point is that the localisation not only takes places in pixel space but also in harmonic space. While in the ILC case the modes at higher ℓ get a very sub-optimal weighting as they do not contribute significantly to the total variance of the one degree map, these modes are weighted much more appropriate in the NILC map. The approach allows to favour foreground rejection on large scales, where foregrounds dominate the total error, and noise rejection on small scales, where foregrounds are negligible but the relative noise level between the various WMAP channels significantly varies. In summary, the NILC map offers a better rejection of Galactic foregrounds than the ILC map and can be considered as the most precise full sky CMB temperature map of the according data release [39].

References

1. G. Gamow, Phys. Rev. **70**, 572 (1946)
2. R. Alpher, R. Herman, Rev. Mod. Phys. **22**, 153 (1950)
3. A. Penzias, R.W. Wilson, ApJ **142**, 419 (1965)
4. G.F. Smoot, C.L. Bennett, A. Kogut, J. Aymon, C. Backus, G. De Amici, K. Galuk, P.D. Jackson, P. Keegstra, L. Rokke, L. Tenorio, S. Torres, S. Gulkis, M.G. Hauser, M.A. Janssen, J.C. Mather, R. Weiss, D.T. Wilkinson, E.L. Wright, N.W. Boggess, E.S. Cheng, T. Kelsall, P. Lubin, S. Meyer, S.H. Moseley, T.L. Murdock, R.A. Shafer, R.F. Silverberg, ApJ Lett. **371**, L1 (1991)
5. J.E. Ruhl, P.A.R. Ade, J.E. Carlstrom, H.M. Cho, T. Crawford, M. Dobbs, C.H. Greer, N.W. Halverson, W.L. Holzapfel, T.M. Lantin, A.T. Lee, J. Leong, E.M. Leitch, W. Lu, M. Lueker, J. Mehl, S.S. Meyer, J.J. Mohr, S. Padin, T. Plagge, C. Pryke, D. Schwan, M.K. Sharp, M.C. Runyan, H. Spieler, Z. Staniszewski, A.A. Stark, Proc. SPIE **5498**, 11 (2004)
6. E.B. Wollack, N. Jarosik, C.B. Netterfield, L. Page, D. Wilkinson, ApJ Lett. **419**, L49 (1993)

7. K. Coble, M. Dragovan, J. Kovac, N.W. Halverson, W.L. Holzapfel, L. Knox, S. Dodelson, K. Ganga, D. Alvarez, J.B. Peterson, G. Griffin, M. Newcomb, K. Miller, S.R. Platt, G. Novak, ApJ Lett. **519**, L5–L8 (1999)

8. C.M. Gutiérrez, R. Rebolo, R.A. Watson, R.D. Davies, A.W. Jones, A.N. Lasenby, ApJ **529**, 47 (2000)

9. J.E. Gallegos, J.F. Macías-Pérez, C.M. Gutiérrez, R. Rebolo, R.A. Watson, R.J. Hoyland, S. Fernández-Cerezo, MNRAS **327**, 1178 (2001)

10. A.C. Taylor, The VSA Collaboration, MNRAS **341**, 1066 (2003)

11. B.P. Crill, P.A.R. Ade, D.R. Artusa, R.S. Bhatia, J.J. Bock, A. Boscaleri, P. Cardoni, S.E. Church, K. Coble, P. deBernardis, G. deTroia, P. Farese, K.M. Ganga, M. Giacometti, C.V. Haynes, E. Hivon, V.V. Hristov, A. Iacoangeli, W.C. Jones, A.E. Lange, L. Martinis, S. Masi, P.V. Mason, P.D. Mauskopf, L. Miglio, T. Montroy, C.B. Netterfield, C.G. Paine, E. Pascale, F. Piacentini, G. Polenta, F. Pongetti, G. Romeo, J.E. Ruhl, F. Scaramuzzi, D. Sforna, A.D. Turner. ApJS **148**, 527 (2003)

12. C.L. Bennett, M. Halpern, G. Hinshaw, N. Jarosik, A. Kogut, M. Limon, S.S. Meyer, L. Page, D.N. Spergel, G.S. Tucker, E. Wollack, E.L. Wright, C. Barnes, M.R. Greason, R.S. Hill, E. Komatsu, M.R. Nolta, N. Odegard, H.V. Peiris, L. Verde, J.L. Weiland, ApJS **148**, 1 (2003)

13. M.C. Runyan, P.A.R. Ade, R.S. Bhatia, J.J. Bock, M.D. Daub, J.H. Goldstein, C.V. Haynes, W.L. Holzapfel, C.L. Kuo, A.E. Lange, J. Leong, M. Lueker, M. Newcomb, J.B. Peterson, J. Ruhl, G. Sirbi, E. Torbet, C. Tucker, A.D. Turner, D. Woolsey, ApJS **149**, 265 (2003)

14. S. Padin, J.K. Cartwright, B.S. Mason, T.J. Pearson, A.C.S. Readhead, M.C. Shepherd, J. Sievers, P.S. Udomprasert, W.L. Holzapfel, S.T. Myers, J.E. Carlstrom, E.M. Leitch, M. Joy, L. Bronfman, J. May, ApJ Lett. **549**, L1 (2001)

15. Tauber et al., A&A, **520**, A1 (2010)

16. Wilkinson Microwave Anisotropy Probe (WMAP): *Seventh Year Explanatory Supplement*, ed. by M. Limon, et al. (NASA/GSFC, Greenbelt, MD, 2011)

17. http://lambda.gsfc.nasa.gov/

18. L. Page, C. Jackson, C. Barnes, C. Bennett, M. Halpern, G. Hinshaw, N. Jarosik, A. Kogut, M. Limon, S.S. Meyer, D.N. Spergel, G.S. Tucker, D.T. Wilkinson, E. Wollack, E.L. Wright, ApJ **585**, 566 (2003)

19. K.M. Górski, E. Hivon, A.J. Banday, B.D. Wandelt, F.K. Hansen, M. Reinecke, M. Bartelmann, ApJ **622**, 759 (2005)

20. http://healpix.jpl.nasa.gov

21. P. Vielva, E. Martínes-González, R.B. Barreiro, J.L. Sanz, L. Cayón, ApJ **609**, 22 (2004)

22. H.K. Eriksen, A.J. Banday, K.M. Gorski, F.K. Hansen, P.B. Lilje, ApJ Lett. **660**, L81 (2007)

23. D. Pietrobon, A. Amblard, A. Balbi, P. Cabella, A. Cooray, D. Marinucci, Phys. Rev. D **78**, 103504 (2008)

24. B. Gold, N. Odegard, J.L. Weiland, R.S. Hill, A. Kogut, C.L. Bennett, G. Hinshaw, X. Chen, J. Dunkley, M. Halpern, N. Jarosik, E. Komatsu, D. Larson, M. Limon, S.S. Meyer, M.R. Nolta, L. Page, K.M. Smith, D.N. Spergel, G.S. Tucker, E. Wollack, E.L. Wright, ApJS **192**, 15 (2011)

25. D. Samtleben, S. Staggs, B. Winstein, Ann. Rev. Nucl. Part. Sci. **57**, 245 (2007)

26. D.P. Finkbeiner, ApJ **146**, 407 (2003)

27. B.T. Draine, A. Lazarian, ApJ Lett. **494**, L19 (1998)

28. B.T. Draine, A. Lazarian, ApJ **508**, 157 (1998)

29. B.T. Draine, A. Lazarian, ApJ **512**, 740 (1999)

30. D.P. Finkbeiner, M. Davis, D.J. Schlegel, ApJ **524**, 867 (1999)

31. N. Jarosik, C.L. Bennett, J. Dunkley, B. Gold, M.R. Greason, M. Halpern, R.S. Hill, G. Hinshaw, A. Kogut, E. Komatsu, D. Larson, M. Limon, S.S. Meyer, M.R. Nolta, N. Odegard, L. Page, K.M. Smith, D.N. Spergel, G.S. Tucker, J.L. Weiland, E. Wollack, E.L. Wright, ApJS **192**, 14 (2011)

32. H. Liu, T.-P. Li, Chin. Sci. Bull. **55**, 907 (2010)

33. H. Liu, T.-P. Li, Chin. Sci. Bull. **56**, 29 (2011)

34. H. Liu, S.-L. Xiong, T.-P. Li, MNRAS Lett., http://arxiv.org/abs/1009.2701v2

35. B.F. Roukema, A&A **518**, A34 (2010)

36. B.F. Roukema, A&A, http://arxiv.org/abs/1007.5307v2
37. G. Hinshaw, M.R. Nolta, C.L. Bennett, R. Bean, O. Dor, M.R. Greason, M. Halpern, R.S. Hill, N. Jarosik, A. Kogut, E. Komatsu, M. Limon, N. Odegard, S.S. Meyer, L. Page, H.V. Peiris, D.N. Spergel, G.S. Tucker, L. Verde, J.L. Weiland, E. Wollack, E.L. Wright, ApJS **170**, 288 (2007)
38. L. Page, G. Hinshaw, E. Komatsu, M.R. Nolta, D.N. Spergel, C.L. Bennett, C. Barnes, R. Bean, O. Dor, J. Dunkley, M. Halpern, R.S. Hill, N. Jarosik, A. Kogut, M. Limon, S.S. Meyer, N. Odegard, H.V. Peiris, G.S. Tucker, L. Verde, J.L. Weiland, E. Wollack, E.L. Wright, ApJS **170**, 335 (2007)
39. J. Delabrouille, J.-F. Cardoso, M. Le Jeune, M. Betoule, G. Fay, F. Guilloux, A&A **493**, 835 (2009)

Chapter 4
Scaling Indices Applied to the WMAP 5-Year Data

4.1 Introduction

The *Wilkinson Microwave Anisotropy Probe (WMAP)* satellite, launched in June 2001, measures the temperature anisotropy of the cosmic microwave background (CMB) radiation with surpassing accuracy, hence providing the best insight on the beginnings of our universe until now. From the first data release on, many investigations were made concerning the Gaussianity of the CMB, since such analyses give information about the nature of the primordial density fluctuations, which are the seeds of those temperature anisotropies. The statistical properties of the density fluctuations are again an important observable for testing cosmological models, especially models of inflation. Standard inflationary models predict the temperature fluctuations of the CMB to be a Gaussian random field which is isotropic and homogenous [1–3]. Still, there also exist more complex models that allow non-Gaussianity in a scale-independent [4–10] or in a scale-dependent way [11, 12]. For a detailed overview of the different models and a more specific survey on scale-dependent ones, see [13] and [14], respectively, as well as enclosed references. In addition, topological defects like cosmic strings can induce local non-Gaussianities and influence the power spectrum [15–18]. Considering this plethora of possible physical mechanisms, which may induce non-Gaussianity, studies of Gaussianity of the CMB are strongly required for testing predictions of fundamental physical theories. By comparing the results with theoretical predictions, we can evaluate which model e.g. of inflation can be accepted or rejected.

Non-Gaussianity implies the presence of any higher order correlations. Therefore, a concrete description of the characteristics of non-Gaussianity is not possible, and one has to state that it can occur in various forms. One can carry out a global analysis to search for deviations from Gaussianity [19–27]. But one can also concentrate on

---

Original publication: G. Rossmanith, C. Räth, A. J. Banday and G. Morfill, *Non-Gaussian Signatures in the five-year WMAP data as identified with isotropic scaling indices*, MNRAS, **399**, 1921 (2009).

G. Rossmanith, *Non-linear Data Analysis on the Sphere*, Springer Theses, DOI: 10.1007/978-3-319-00309-2_4, © Springer International Publishing Switzerland 2013

more specific investigations (this is most often performed in addition to a general analysis), whereas we want to point out the following two:

Investigations concerning *asymmetries* in the CMB data were accomplished with linear [28–34] as well as non-linear methods [28, 35–41]. With those methods, studies of the differences between the northern and southern hemisphere of the galactic coordinate system, naturally given by the absent region of the outmasked galactic plane, as well as a search for a preferred direction of maximum asymmetry were performed. In almost all investigations, significant asymmetries between the north and the south were detected. Thereby, it depended on the type of analysis, which hemisphere featured the larger deviations from Gaussianity and which hemisphere agreed better with the standard model. The preferred direction of maximal disparity was in most investigations found to lie close to the ecliptic axis.

Local features are another particular form of non-Gaussianity being of growing importance, e.g. for the search of topological defects like cosmic strings. Since the first detection of the famous *cold spot* by [37], many investigations tried to find new, or re-detect known spots with various methods [40, 42–51]. In doing so, several significant spots have been detected up to now.

In [52], all mentioned investigations were accomplished by applying, for the first time, the scaling index method on the WMAP 3-year data. In this paper, we continue these analyses by applying the scaling index method on the WMAP 5-year observations. We search for global non-Gaussianities and asymmetries in the data and use a modified approach to detect local features.

This chapter is structured as follows: In Sect. 4.2 we present the preprocessing of the WMAP data and the modality of creating the simulations. In Sect. 4.3, the scaling index method is introduced as well as a technique to cope with boundary effects. With these requisites, we are ready to perform our calculations, whose results are presented in Sect. 4.4. In this chapter, we first discuss the global investigations as well as asymmetries and focus on local features later on. All these findings are summarised in Sect. 4.5. Finally, we draw our conclusions in Sect. 4.6.

4.2 WMAP Data and Simulations

For our investigations we use the Q-, V- and W-band five-year-data of the WMAP-satellite as it is provided by the WMAP-Team.[1] We work with the foreground-reduced maps, which use the Foreground Template Model proposed in [53] and [54] for foreground reduction. To obtain a co-added VW-map as well as single V-, W- and Q-maps, we accumulate the differencing assemblies Q1, Q2, V1, V2, W1, W2, W3, W4 via a noise-weighted sum [55]:

$$T(\theta, \phi) = \frac{\sum_{i \in \mathcal{A}} T_i(\theta, \phi)/\sigma_{0,i}^2}{\sum_{i \in \mathcal{A}} 1/\sigma_{0,i}^2} \qquad (4.1)$$

---

[1] http://lambda.gsfc.nasa.gov

In this equation, $\mathcal{A}$ characterises the set of required assemblies, e.g. for the co-added VW-map $\mathcal{A} = \{V1, V2, W1, W2, W3, W4\}$. The parameters θ and ϕ correspond to the co-latitude and the longitude on the sphere, while the five-year noise per observation of the different assemblies, given by [56], is denoted by σ_0.

We decrease the resolution of the maps to 786432 pixels, which equals to $N_{side} = 256$ in the employed HEALPix-software[2] [57] and cut out the heavily foreground-affected parts of the sky using the KQ75-mask [58], which has to be downgraded as well. We choose a conservative downgrading of the mask by taking only all pixels at $N_{side} = 256$ that do completely consist of non-mask-pixels at $N_{side} = 512$. All downgraded pixels at $N_{side} = 256$, for which one or more pixels at $N_{side} = 512$ belonged to the KQ75-mask, are considered to be part of the downgraded mask as well. In doing so, 28.4 % of the sky is removed (see upper left part of Fig. 4.1). Finally, we remove the residual monopol and dipol by means of the appropriate HEALPix routine applied to the unmasked pixels only.

To accomplish a test of non-Gaussianity, we also need simulations of Gaussian random fields. We create 1000 simulations for every band and proceed hereby as follows: We take the best fit ΛCDM power spectrum C_l, derived from the WMAP 5-year data only, and the respective window function for each differencing assembly (Q1-Q2, V1-V2, W1-W4), as again made available on the LAMBDA-website (see footnote 1). With these requisites, we can create Gaussian random fields mimicing

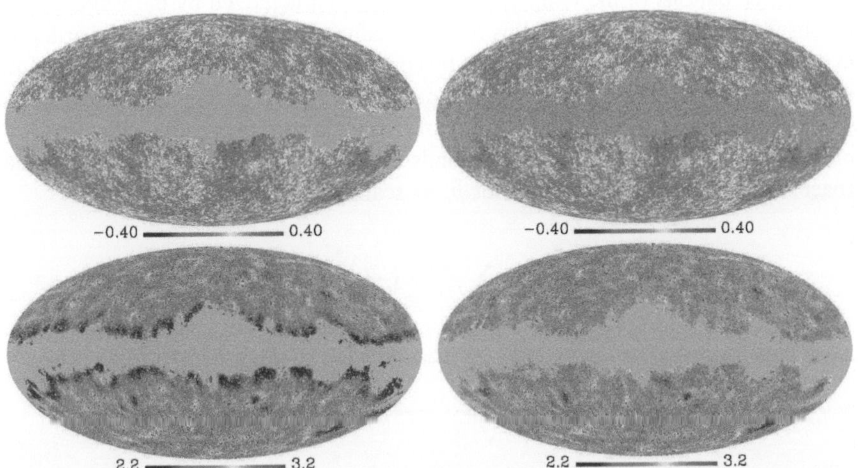

Fig. 4.1 The two plots on the *left hand side* illustrate the original 5-year WMAP-map of the co-added VW-band (*above*) and the related colour-coded α-response (*below*). The equivalent plots for the mask-filling technique are arranged on the *right hand side*. These maps (and all following ones) are shown in a conventional scheme, namely the Mollweide projection in the Galactic reference frame with the Galactic Centre at the centre of the image and the longitude increasing from there to the *left-hand side*

---

[2] http://healpix.jpl.nasa.gov

the Gaussian properties of the best fit ΛCDM-model and including the WMAP-specific beam properties by convolving the C_l's with the window function. For every assembly, we add Gaussian noise to these maps with a particular variance for every pixel of the sphere. This variance depends on the number of observations $N_i(\theta, \phi)$ in the respective direction and the noise dispersion per observation, $\sigma_{0,i}$. After this procedure, we summarize the Q-, V- and W-bands and the co-added VW-band using Eq. (4.1), decrease the resolution to $N_{side} = 256$, cut out the KQ75-mask and remove the residual monopol and dipol, just as we did with the WMAP-data.

4.3 Weighted Scaling Index Method

4.3.1 Formalism

We perform our investigations using the scaling index method (SIM) [59, 60], which enables a characterisation of the structure of a given data set. It has already been used in time series analysis of active galactic nuclei (AGN) [61, 62] as well as in structure analysis for 2D and 3D image data, e.g. in [63–65]. In the following, we only present a short overview of the calculation of scaling indices. For a more detailed formalism of using the SIM in CMB analysis we refer to [52].

The fluctuations of the temperature maps are characterized by the values of the pixelised sky of a sphere with radius R. We transform this representation to variations in the radial direction around the sphere by applying a jitter depending on the intensity of the fluctuation. Thereby we obtain a point-distribution in the three-dimensional space. Thus, given N_{pix} as the number of pixels on the sphere, the value of every pixel (θ_i, ϕ_i), $i = 1, ..., N_{pix}$ corresponds to a vector $\vec{p}_i$ in the three-dimensional space. We then define for every point $\vec{p}_i$ its scaling index by

$$\alpha(\vec{p}_i, r) = \frac{\sum_{j=1}^{N_{pix}} 2 \left(\frac{d_{ij}}{r}\right)^2 e^{-\left(\frac{d_{ij}}{r}\right)^2}}{\sum_{j=1}^{N_{pix}} e^{-\left(\frac{d_{ij}}{r}\right)^2}} \tag{4.2}$$

where d_{ij} denotes the euclidian distance measure

$$d_{ij} = \|\vec{p}_i - \vec{p}_j\|_2$$

between the points $\vec{p}_i$ and $\vec{p}_j$, while r characterizes a scale parameter. This parameter does not draw a clear-cut line between the pixels that are included in the calculations and those that are excluded; it rather influences how each single pixel is taken into consideration for the calculation, in relation to its distance from the centre pixel: For lower r, only the closest pixels are important in the calculation of $\alpha(\vec{p}_i, r)$, whereas for larger r, the farther distant pixels are considered as well, even though

Table 4.1 The angular scales corresponding to the position of the 90% (ℓ_1) and the 10% (ℓ_2) weighting in the scaling index formula when using a given scale parameter r

Radius	0.025	0.050	0.075	0.100	0.125
$[\ell_1, \ell_2]$	[83,387]	[41,193]	[28,129]	[21,97]	[17,77]
Radius	0.150	0.175	0.200	0.225	0.250
$[\ell_1, \ell_2]$	[14,65]	[12,55]	[10,48]	[9,43]	[8,39]

with a still lower weight than the close pixels. In our study, we use the ten scales $r_i = 0.025, 0.05, ..., 0.25$, $i = 1, 2, ..., 10$ and the radius $R = 2$ for the sphere. Table 4.1 shows for each radius r the corresponding angular scales at the position of the 90% and the 10% weighting, thus giving an estimate on how the r-values relate to ℓ-bands in Fourier space.

The value of α characterises the structural components of a point distribution. For example, points in a cluster-like, filamentary or sheet-like structure lead to $\alpha \approx 0$, $\alpha \approx 1$ or $\alpha \approx 2$, respectively. A uniform distribution of points results in $\alpha \approx 3$, while points in underdense regions in the vicinity of point-like structures, filaments or walls have $\alpha > 3$.

On the basis of these scale-dependent α-values, we compute simple measures such as moments and empirical probability distributions. We make use of the following scale-dependent statistics, namely the mean, the standard deviation and a diagonal χ^2-statistics, to compare the results of the original WMAP data with the results of the simulations:

$$\langle \alpha(r_k) \rangle = \frac{1}{N} \sum_{j=1}^{N} \alpha(\vec{p}_j, r_k) \tag{4.3}$$

$$\sigma_{\alpha(r_k)} = \left(\frac{1}{N-1} \sum_{j=1}^{N} \left[\alpha(\vec{p}_j, r_k) - \langle \alpha(r_k) \rangle \right]^2 \right)^{1/2} \tag{4.4}$$

$$\chi^2_{\alpha(r_k)} = \sum_{i=1}^{2} \left[\frac{M_i(r_k) - \langle M_i(r_k) \rangle}{\sigma_{M_i(r_k)}} \right]^2 \tag{4.5}$$

where $M_1(r_k) = \langle \alpha(r_k) \rangle$, $M_2(r_k) = \sigma_{\alpha(r_k)}$ and N denotes the number of pixels in consideration. For all analyses we will only consider the non-masked pixels of the full sky or of (rotated) hemispheres, as it will be outlined in Sect. 4.4.1. Note that we follow the reasoning of [35] and choose a diagonal and not the full χ^2-statistics involving the inverted cross-correlation matrix, because also in our case the moments are highly correlated leading to high values in the off-diagonal elements of the cross-correlation matrix. Therefore, the matrix would converge very slowly and numerical stability would not be given. If however the chosen model is a proper description of the data, *any* combination of measures should yield statistically the same values for the observations and the simulations.

To obtain scale-independent variables as well, we also use three diagonal χ^2-statistics, derived from $\langle\alpha\rangle$ and σ_α, which sum over all utilised length scales r:

$$\chi^2_{\langle\alpha\rangle} = \sum_{k=1}^{N_r}\left[\frac{M_1(r_k) - \langle M_1(r_k)\rangle}{\sigma_{M_1(r_k)}}\right]^2 \tag{4.6}$$

$$\chi^2_{\sigma_\alpha} = \sum_{k=1}^{N_r}\left[\frac{M_2(r_k) - \langle M_2(r_k)\rangle}{\sigma_{M_2(r_k)}}\right]^2 \tag{4.7}$$

$$\chi^2_{\langle\alpha\rangle,\sigma_\alpha} = \sum_{i=1}^{2}\sum_{k=1}^{N_r}\left[\frac{M_i(r_k) - \langle M_i(r_k)\rangle}{\sigma_{M_i(r_k)}}\right]^2 \tag{4.8}$$

Hereby denotes $M_1(r_k) = \langle\alpha(r_k)\rangle$ and $M_2(r_k) = \sigma_{\alpha(r_k)}$. The number of different scale parameters r is named N_r. Throughout all subsequent investigations, N_r equals ten.

Finally, to be able to access the degree of difference between the data and the simulations and hence a degree of the non-Gaussianity of the data, we use the σ-normalised deviation of the WMAP results of the above-mentioned statistics:

$$S = \frac{M - \langle M\rangle}{\sigma_M} \tag{4.9}$$

where in this case M refers to one of the variables defined in Eqs. (4.3)–(4.8) respectively. M itself is calculated by using the WMAP data, while its moments result from the simulations. Note that we pass on the absolute value in this general definition to obtain positive as well as negative deviation. Although we will use the absolute value in the global investigations, the sectioning into positive and negative deviation is useful for the analysis of north–south asymmetry by means of rotated hemispheres in Sect. 4.4.1. It also allows a better interpretation of the character of difference, since e.g. a higher mean of the scaling indices implies a more 'unstructured' arrangement of the 'pixel cloud' and vice versa. Similarly, a higher standard deviation of the indices indicates a larger structural variability.

In the tables, we also included the fraction p of the simulations that have higher (lower) values than the data in terms of the respective calculated statistics. This percentage corresponds to a empirical significance level of the null hypothesis that the observation belongs to a Gaussian Monte Carlo ensemble.

4.3.2 Coping with Boundary Effects

The regions in the direction of the galactic plane as well as many small spots all over the WMAP map are masked out since they represent heavily foreground-affected areas which would not allow a reasonable analysis of the intrinsic background

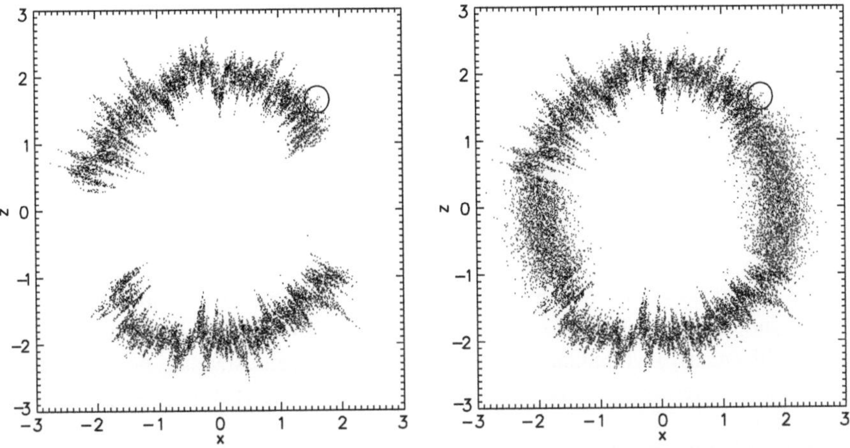

Fig. 4.2 A slice of the three-dimensional representation of the VW-band WMAP data, illustrated as a *x,z*-projection of all points with $|y| < 0.05$. The plot on the *left* illustrates the original, the one on the right the mask-filling method. The *black circles* indicate the scaling range $r = 0.2$

fluctuations. But this operation spoils the results of the scaling index method: Instead of a more or less uniform distribution, the α-values in the regions around the mask now detect a sharp boundary with no points in the masked area, into which the scaling regions extends (see Fig. 4.2). This results in lower values of α. The effect can clearly be seen in the α-response of the masked VW-band WMAP-data in the lower left corner of Fig. 4.1. A solution to this problem is to *fill* the masked areas with suitable values, that prevent the low outcome at the edges of the mask. We accomplish this by filling in (nearly) white Gaussian noise with adjusted parameters. This is performed by applying the following two steps:

At first, we fill the masked regions with Gaussian noise, whose standard deviation for each pixel corresponds to the pixel noise made available on the LAMBDA-website[3]:

$$T^*_{mask}(\theta, \phi) \sim \mathcal{N}(0, \sigma^2_{(\theta,\phi)})$$

Here, $\sigma_{(\theta,\phi)}$ denotes the pixel noise of the pixel which is located in the direction (θ, ϕ). Then, we scale the expectation value and the variance as a whole to the empirical mean μ_{rem} and variance σ^2_{rem} of the remaining regions of the original temperature map:

$$T_{mask}(\theta, \phi) = \frac{\sigma^2_{rem}}{\sigma^2_{mask}}\, T^*_{mask}(\theta, \phi) + \mu_{rem}$$

with

---

[3] http://lambda.gsfc.nasa.gov

$$\mu_{rem} = \frac{1}{N_{\mathcal{R}}} \sum_{(\theta,\phi)\in\mathcal{R}} T(\theta,\phi)$$

$$\sigma^2_{rem} = \frac{1}{N_{\mathcal{R}}-1} \sum_{(\theta,\phi)\in\mathcal{R}} (T(\theta,\phi) - \mu_{rem})^2$$

$$\sigma^2_{mask} = \frac{1}{N_{\mathcal{M}}-1} \sum_{(\theta,\phi)\in\mathcal{M}} T^*_{mask}(\theta,\phi)^2$$

where $\mathcal{R}$ and $\mathcal{M}$ stand for the non-masked and masked region of the map respectively, and $N_{\mathcal{R}}$ as well as $N_{\mathcal{M}}$ denote their number of pixels. Thus, we filled the mask with (nearly) white Gaussian noise whose mean and standard deviation equal the respective terms of the remaining map, whereby the spatial noise patterns are preserved.

With this filling technique, we obtain a complemented data set instead of just excluding the masked regions. Figure 4.2 shows a slice of the three-dimensional representation of the temperature fluctuations of both techniques, representing a 2D projection of the 3D point distribution used for the calculation of the scaling indices. The centre region of the filled sphere now highly resembles the appearance of the remaining area, although a more uniform arrangement is visible. In the form of a mollweide projection, the filling method as well as the corresponding α-response are displayed in the right column of Fig. 4.1.

The filling strategy shows obvious success in the adjustment of the scaling indices map (see the lower right panel of Fig. 4.1): The white noise leads to higher values in the α-response for the pixels close to the mask as compared to the masking method. The regions around the edges of the mask feature now α-responses that match far better the values of the remaining regions. Still these α-values are calculated with the help of an artificial environment, but now the contortions are lower compared to the original approach. Since we apply this method to both the original WMAP data as well as to the simulations, the now smaller systematic errors in the α-calculation for the 'edge'-pixels are the same for both kinds of maps. Thus any significant deviation found in the WMAP data is due to intrinsic effects.

The most important advantage of the filling strategy arises if one considers local features in the CMB map: For a search of local anomalies (e.g. cold spots), the filled map provides a better underlying than the original map: In a point distribution, spots as well as points at the border of the distribution show similarly low α-values. Thus, by cutting out the mask, it is difficult to decide whether a detected local feature really exists or whether it originates from a masked area nearby. But by using the filling technique, there is no longer an edge between the masked and the non-masked regions, and anomalies in the Gaussian noise of the mask are highly improbable. Thus, any detected local feature must then originate from the data itself. Considering the amount of masked areas *outside* the galactic plane, this technique describes a considerable improvement for the whole sky. Due to these advantages we will only use the mask-filled maps in Sect. 4.4.2.

4.4 Results

4.4.1 Band-wise and Co-added Map Analysis

In Fig. 4.3, the empirical probability densities $P(\alpha)$ of the scaling indices (calculated with $r = 0.2$) are displayed for the WMAP data and for the simulations, evaluated for the original and the filling method from Sect. 4.3.2. For clarity reasons we only used the first 50 simulations in these plots. For both methods, a shift of the WMAP data to higher values can be detected, that becomes particularly apparent in the northern hemisphere of the galactic coordinate system. This indicates a more 'unstructured' arrangement of the underlying temperature fluctuations of the CMB data in comparison to the simulations. In addition, the histograms of the simulations are slightly broader and therewith containing a larger structural variability than the one of the WMAP data.

Comparing the non-filling and the filling method, the histograms of the latter feature a higher maximum as well as higher values for large α, but lower probabilities for $\alpha \in [2.0, 2.5]$. The obvious reason for this shift is the fact that the filled mask does not reduce the α-values of its surroundings as it was the case with the former method. Now, the outcome of these regions is influenced by the white noise and is therefore allocated at higher values.

If we focus on the mean values $\langle \alpha(r_k) \rangle$ of the scaling indices, and compare the results of the simulations with the WMAP data, the above mentioned shift to higher values becomes yet clearer. This can be seen in Fig. 4.4, where the distribution of $\langle \alpha(r_k) \rangle$ for the simulations as well as the data is displayed for the five different scale parameters r_2, r_4, r_6, r_8 and r_{10}. These results were obtained using the full sky. For all applied scales, the distance between the average over all simulations and the result of the original WMAP map is notably similar. If we perform this analysis for the northern hemisphere only, the deviations of the original data as compared to the simulations become significantly larger.

Both the shift to higher values of the WMAP data in comparison to the simulations as well as its broader density are reflected in the σ-normalised deviations $S(r)$ of the scale-dependent statistics of the Eqs. (4.3)–(4.5): The former is reflected in the mean and the latter in the standard deviation (and therefore both aspects in the diagonal χ^2-statistics). Figure 4.5 shows these deviations for the coadded VW-band as a function of the scale parameter r for the original and the mask filling method, while Fig. 4.6 displays only the latter method, but for the three single bands. As above, the results are illustrated for the full sky as well as for the separate hemispheres. The shift to higher values of the WMAP data in the northern hemisphere in Fig. 4.3 appears now as an increased $S(r)$, especially for higher scales ($r \geq 0.125$), where the deviations of the two moments range between 2σ and 3.5σ, and the χ^2-combination nearly reaches a 6σ-level. In the southern hemisphere, only the lowest scales show a namable $S(r)$. On larger scales, no signatures for deviations from Gaussianity are identified. Looking at the single bands Q, V and W, the overall qualitative behaviour of the images is quite similar, while the σ-normalised deviations itself are slightly lower in most cases.

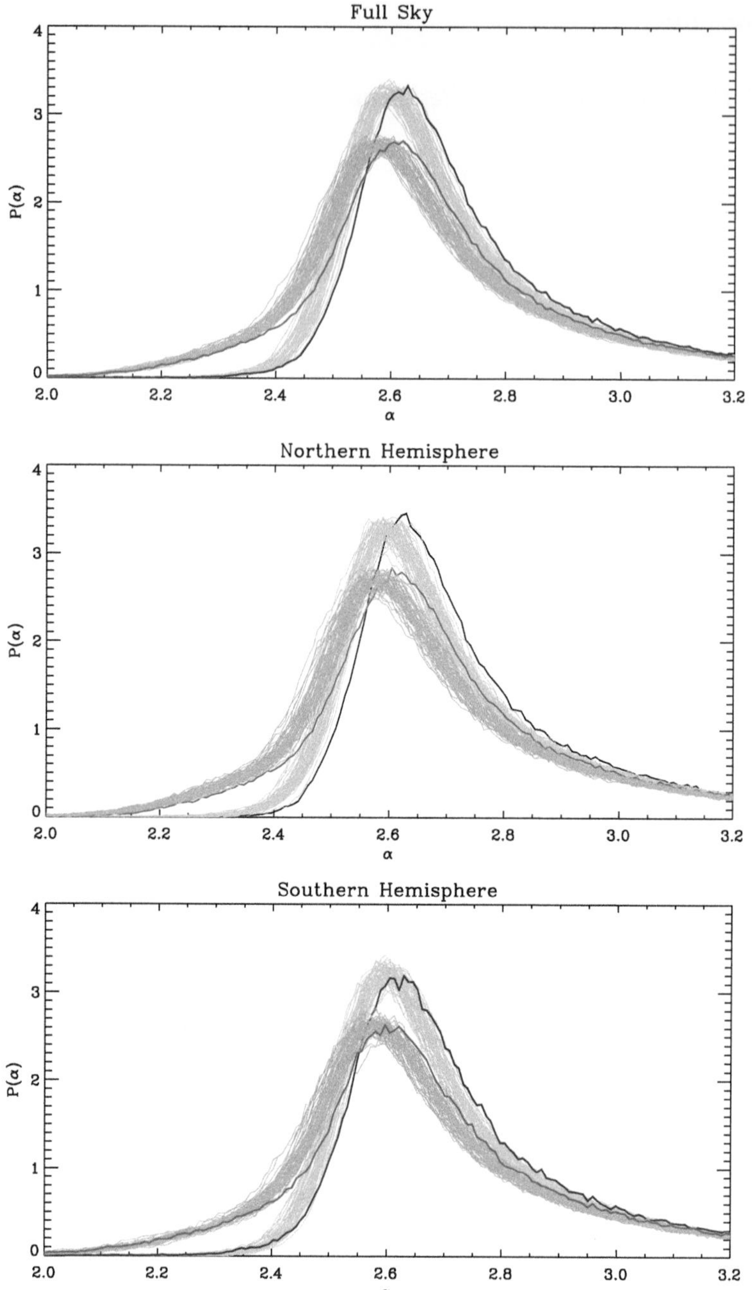

Fig. 4.3 The probability distributions $P(\alpha)$ of the scaling indices for the WMAP data (*dark lines*) and for 50 simulations (*fainter lines*) by using the scale parameter $r = 0.2$, computed for the original (*red*) and the mask filling method (*blue*). The *upper* histogram shows the distribution of the full sky data set, while the *middle* and the *lower* ones show the distribution of the northern and southern hemisphere respectively

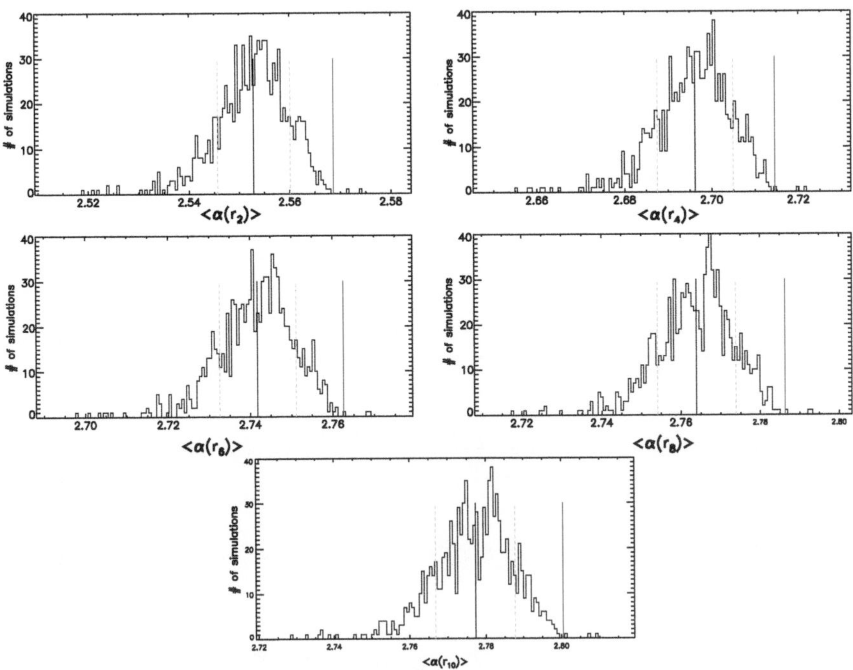

Fig. 4.4 The histograms of the mean values $\langle\alpha(r_k)\rangle$ of the 1000 simulations for the five different scale parameters r_2, r_4, r_6, r_8 and r_{10} (starting from *upper left*), calculated for the full sky. The *red lines* denote the corresponding results of the WMAP data, while the *black* and *grey lines* characterise the average over all simulations and the 1σ regions, respectively

A remarkable fact is the appearance of the *highest* $S(r)$ (5.2σ for the χ^2-combination in the northern hemisphere at $r = 0.15$) in the frequency band which is considered to be the *least* foreground-contaminated one, namely the V-band. Comparing the co-added VW-band of the original approach and of the mask filling method, the σ-normalised deviations of the mean are almost identical. The standard deviation of the latter method in comparison to the former one shows a slightly lower $S(r)$ for higher scales, which is also reflected in the graph of the χ^2-combination, yet the profile remains the same.

We also calculated the σ-normalised deviations S and the percentages p of the simulations with higher (lower) results of the scale-independent diagonal χ^2-statistics from the Eqs. (4.6) to (4.8), which are listed in Table 4.2. Although the results are damped by a few unimportant scales, high deviations are still found, particularly in the northern hemisphere. For a better comparison to separate scale lengths, the respective results of the scale-dependent statistics (4.3)–(4.5) are listed in Table 4.3, for which we used the single scale $r = 0.2$.

In general, all occurring characteristics of the Figs. 4.3 and 4.5 match the findings of the analysis of the WMAP 3-year data in [52]. This indicates that the results are not based on some time-dependent effects. Since the 5-year data features lower error

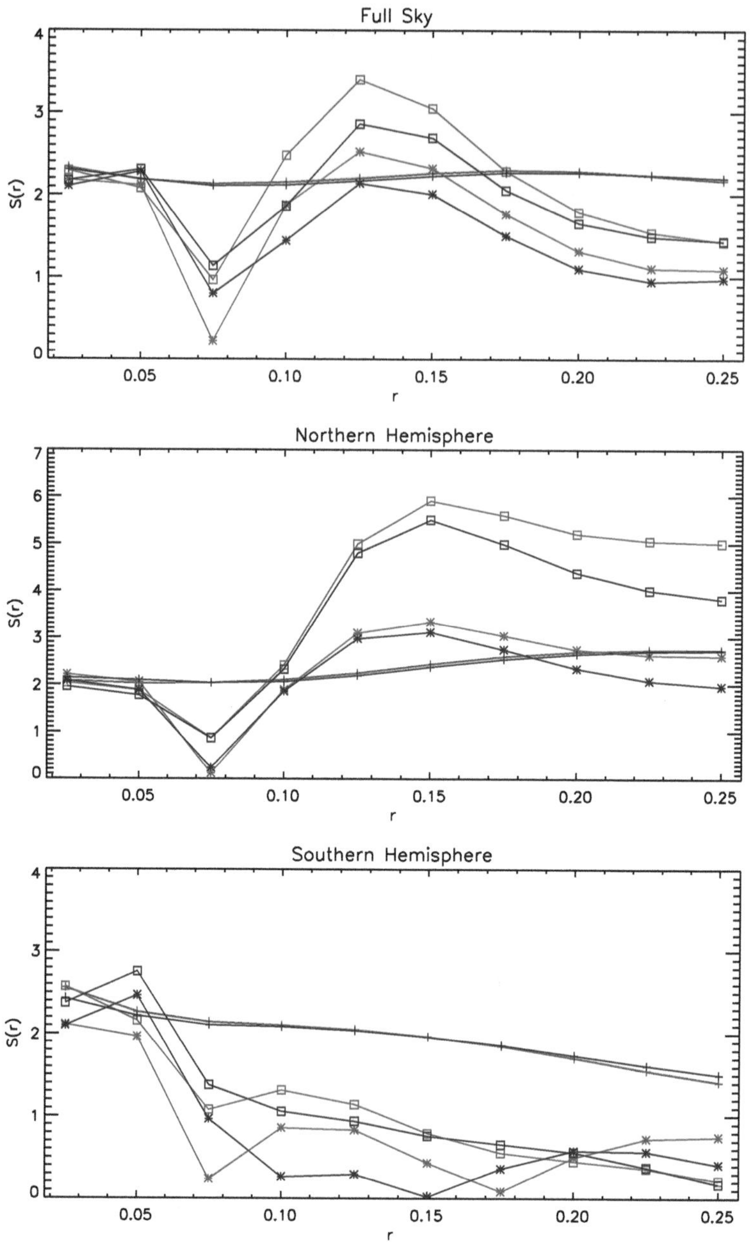

Fig. 4.5 The σ-normalised deviations $S(r)$ of the statistics of the Eqs. (4.3)–(4.5) in absolute values for the VW-band, plotted as a function of the scale parameter r. The lines with "+" denote the mean, "*" the standard deviation and the boxes the χ^2-combination, each for the original (*red*) and the mask filling method (*blue*). As in Fig. 4.3, the *upper* diagram shows the results of a full-sky analysis, while the middle and the *lower* ones show the results when only concerning the northern or southern hemisphere respectively

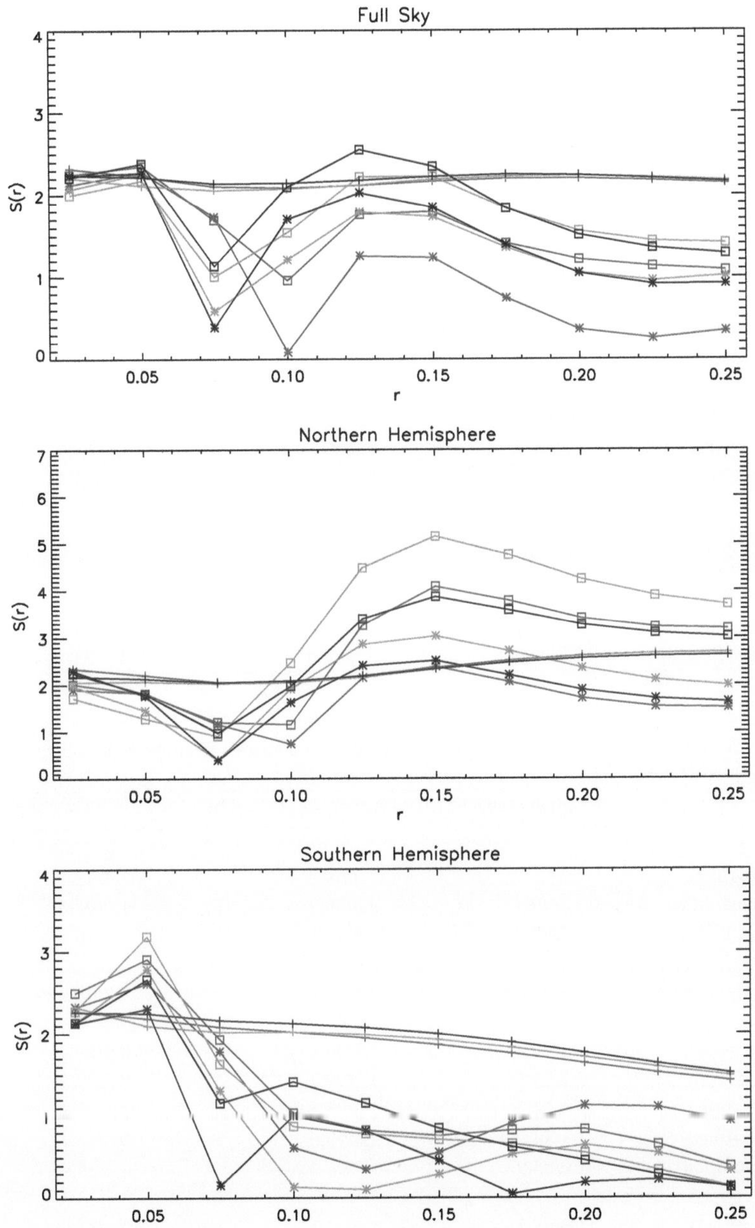

Fig. 4.6 Same as Fig. 4.5 but applied to the Q- (*red*), V- (*green*) and W-band (*blue*). Only the results of the mask-filling method are shown while the original method is left out

Table 4.2 The σ-normalised deviations S and the empirical probabilities p of the scale-independent diagonal χ^2-statistics from the Eqs. (4.6) to (4.8) for the different bands and methods as well as for the Full Sky and the single hemispheres

	Full sky ($S/\%$)	Northern sky ($S/\%$)	Southern sky ($S/\%$)
$\chi^2_{\langle\alpha\rangle}$			
VW (original)	2.2/97.3	2.7/98.5	1.7/96.6
VW (mask-filled)	2.2/97.3	2.7/98.3	1.7/96.6
Q (mask-filled)	2.1/97.4	2.7/98.3	1.5/95.8
V (mask-filled)	2.0/97.4	2.6/98.1	1.5/96.2
W (mask-filled)	2.1/97.5	2.6/98.2	1.7/96.6
$\chi^2_{\sigma_\alpha}$			
VW (original)	2.0/95.6	5.5/99.7	0.1/68.5
VW (mask-filled)	1.6/93.4	4.3/99.3	0.3/72.9
Q (mask-filled)	0.7/83.1	2.3/96.0	1.2/89.4
V (mask-filled)	1.2/90.5	4.0/98.9	0.6/79.2
W (mask-filled)	1.4/92.3	2.8/96.7	1.9/71.6
$\chi^2_{\langle\alpha\rangle,\sigma_\alpha}$			
VW (original)	2.3/97.4	4.2/99.1	1.3/93.5
VW (mask-filled)	2.1/97.1	3.7/98.8	1.3/94.2
Q (mask-filled)	1.8/96.3	2.9/98.3	1.6/95.5
V (mask-filled)	1.9/96.6	3.5/98.8	1.3/94.1
W (mask-filled)	2.0/96.4	3.0/98.5	1.3/93.7

Table 4.3 Same as Table 4.2, but for the scale-dependent statistics from the Eqs. (4.3) to (4.5) for the single scale $r = 0.2$

	Full sky ($S/\%$)	Northern sky ($S/\%$)	Southern sky ($S/\%$)
$\langle\alpha(0.2)\rangle$			
VW (original)	2.3/99.7	2.7/99.8	1.7/97.1
VW (mask-filled)	2.3/99.7	2.3/98.7	1.7/97.6
Q (mask-filled)	2.2/99.6	2.6/99.8	1.6/96.6
V (mask-filled)	2.2/99.6	2.6/99.8	1.7/97.2
W (mask-filled)	2.2/99.6	2.6/99.8	1.8/98.2
$\sigma_{\alpha(0.2)}$			
VW (original)	1.3/90.7	2.7/99.8	0.5/68.9
VW (mask-filled)	1.1/85.6	4.3/99.3	0.6/69.8
Q (mask-filled)	0.4/64.2	1.7/95.6	1.1/85.8
V (mask-filled)	1.1/84.9	2.4/99.0	0.6/71.4
W (mask-filled)	1.0/83.9	1.9/96.4	0.2/55.9
$\chi^2_{\alpha(0.2)}$			
VW (original)	1.8/95.3	5.2/99.4	0.5/81.6
VW (mask-filled)	1.7/94.5	4.4/99.1	0.6/82.6
Q (mask-filled)	1.2/90.9	3.4/98.6	0.8/97.1
V (mask-filled)	1.6/94.2	4.2/99.2	0.5/81.9
W (mask-filled)	1.5/93.7	3.3/98.6	0.5/81.0

bars than the 3-year data, it is also improbable that both results are induced by noise effects only.

Evidence for north–south asymmetry in the WMAP data was already detected using the angular power spectrum [29, 30] and higher order correlation functions [28], spherical wavelets [37], local curvature analysis [38], two-dimensional genus measurements [36] as well as all three Minkowski functionals [35], correlated component analysis [31], spherical needlets [40], frequentist analysis of the bispectrum [39], two-point correlation functions [32, 33] and Bayesian analysis of the dipole modulated signal model [34]. To take a closer look at asymmetries in the WMAP five-year data in our investigations, we perform an analysis of rotated hemispheres, as it was done for the three-year data in [52]: For 3072 different angles, we rotate the original and simulated maps and then compute $S(r)$ for the above statistics (mean, standard deviation and χ^2-combination) by only using the data in the resulting new upper hemisphere. Thus, the colour of each pixel in the corresponding Fig. 4.7 expresses the positive or negative σ-normalised deviation $S(r)$ of the hemisphere around that pixel in the WMAP-data compared to the hemispheres around that pixel in the simulations. We apply this analysis for the co-added VW-band as well as for the single bands, whereas for the VW-band we use both the original and the mask filling method, but for the single bands the filling method only. In all charts of Fig. 4.7 we can detect an obvious asymmetry in the data: The largest deviations between the data and the simulations are exclusively obtained for rotations pointing to northern directions relative to the galactic coordinate system. The maximum value for $S(r)$ of the χ^2 analysis (right column of Fig. 4.7) using the mask-filling method on the co-added VW-band is obtained in the reference frame pointing to $(\theta, \phi) = (27°, 35°)$, which is close to the galactic north pole. This proximity to the pole is consistent to the results of [38] and [52], as well as to those findings of [29] and [28] that considers large angular scales. For the standard deviation (central column of Fig. 4.7), the northern and southern hemispheres offer different algebraic signs. The negative $S(r)$ of the north implies a lower variability than the simulations in this region, while the south shows a converse behaviour. The fact that the plots using the new method show slightly lower values for $S(r)$ than the ones using the old method may be explained by the fraction of pure noise values within every rotated hemisphere, that diminish the degree of difference between the data and the simulations.

Another remarkable feature of Fig. 4.7 is the high correlation between the different bands, that is visible to the naked eye but also confirmed mathematically. By calculating correlations c among all combinations of those bands where the mask filling method was applied, we obtain for the mean $c \geq 0.99$ and for the standard deviation as well as the χ^2-statistics $c \geq 0.95$. While the Q-band is heavily foreground-affected, first of all by synchrotron radiation as well as radiation from electron-ion scattering ("free-free emission"), the W-band is mainly distorted by Dust emission. The V-band is affected by all these foregrounds, even though less than the other bands. As mentioned in Sect. 4.2, we use the foreground-reduced maps in our analysis, but one could still expect some small interferences. Despite the different influences on the different bands, we obtain the same signatures of non-Gaussianity in all single bands as well as in the co-added VW-band. Therefore we conclude that

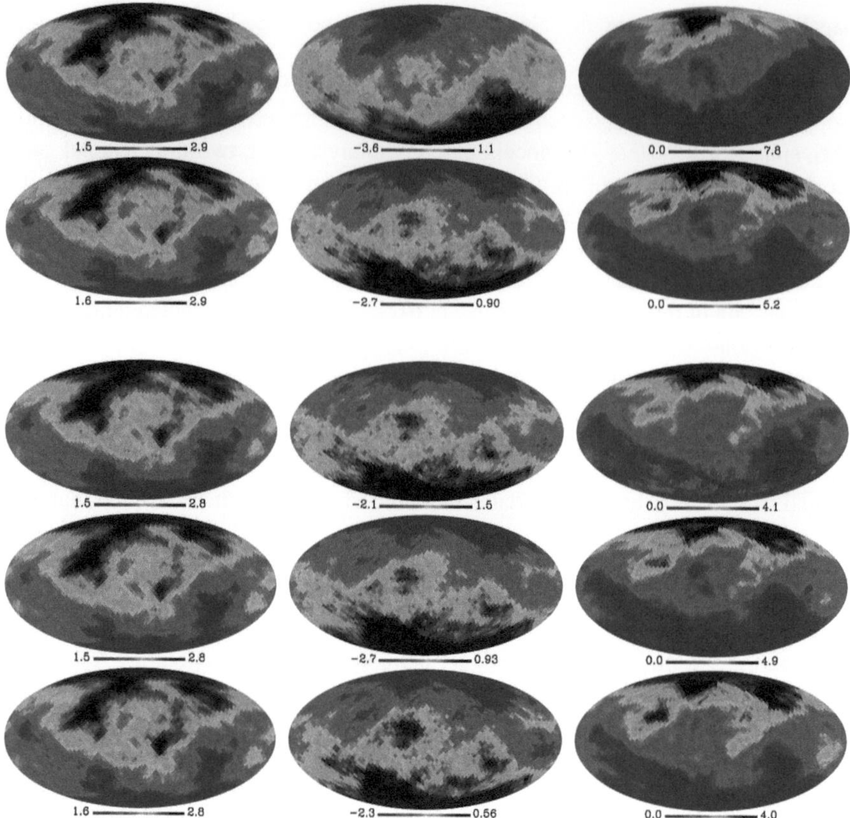

Fig. 4.7 The σ-normalised deviations $S(r)$ of the rotated hemispheres at the scale parameter $r = 0.2$ for the mean (*left column*), the standard deviation (*central column*) and the diagonal χ^2-statistics (*right column*) for the co-added VW-Band without (*top row*) and with (*second row*) the appliance of the mask filling method, as well as for the single Q-, V- and W-bands (*third* to *fifth row*), for which the mask filling method was always applied. Notice the different colour scaling for each plot

the measured asymmetry is very unlikely the result of a foreground influence but has to be concluded of thermal origin.

4.4.2 Local Features

An interesting anomaly in the CMB data is that there are small regions which show very high or very low values in some local structure analysis. Vielva et al. (2004) detected the first of these regions, the well-known *cold spot* at $(\theta, \phi) = (147°, 209°)$ a few years ago by using a wavelet analysis. This Spot was re-detected several times using amongst others wavelet analysis [32, 44–47], scaling indices [52] or the Kolmogorov stochasticity parameter [42]. Furthermore, there have been some investigations which, in addition to the re-detection of the first spot, detected secondary spots via directional [48–50] or steerable wavelets [51], needlets [40] and again the Kolmogorov stochasticity parameter [43]. These spots could be the result of some yet not fully understood physical process. For the *cold spot* lots of theories already exist which try to explain its origin by second-order gravitational effects [66, 67], a finite universe model [68], large dust-filled voids [69–72], cosmic textures [73], non-Gaussian modulation [74], topological defects [75], textures in a brane world model [76] or an asymptotically flat Lemaitre-Tolman-Bondi model [77, 78].

For our investigations concerning spots in the WMAP data we only use the mask-filling method of Sect. 4.3.2 due to the reasons already explained above. We extend the analysis of scaling indices by applying two different approaches to detect anomalies: The first one is to calculate the σ-normalised deviation of *every pixel* on the α-response of the CMB map. For a given scale parameter r, this is achieved by comparing the scaling index $\alpha(\vec{p}_i, r)$ of each vector $\vec{p}_i$, $i = 1, ..., N_{pix}$, of the original data with the mean of the corresponding values $\alpha_\ell(\vec{p}_i, r)$, $\ell = 1, ..., N_{sim}$, of the simulations depending on their standard deviation, where N_{sim} denotes the number of the simulations. Formally, this reads as:

$$S_{i,r} = \frac{\alpha(\vec{p}_i, r) - \mu_{i,r}}{\sigma_{i,r}}, \tag{4.10}$$

with

$$\mu_{i,r} = \frac{1}{N_{sim}} \sum_{\ell=1}^{N_{sim}} \alpha_\ell(\vec{p}_i, r)$$

$$\sigma_{i,r}^2 = \frac{1}{N_{sim} - 1} \sum_{\ell=1}^{N_{sim}} \left(\alpha_\ell(\vec{p}_i, r) - \mu_{i,r} \right)^2$$

The results are illustrated in the upper left part of Fig. 4.8.

The second approach smoothes the α-maps of the original and simulated data by computing for every pixel the mean value of its surroundings given by some specified maximum distance, which equals $3°$ in our analysis. We apply the pixel-wise deviations $S_{i,r}$ again on the resulting maps. The outcome of this procedure is shown in the upper right part of Fig. 4.8. In the lower left plot of the same figure only the deviations $S_{i,r} \leq -3.0$ are illustrated to gain yet another clearer view on the interesting areas.

The first approach clearly shows the *cold spot* and indicates some secondary spots in the southern as well as in the northern hemisphere. These get confirmed in the plot of the smoothing method, where we obtain a deviation of up to -7σ for several clearly visible areas: In the southern hemisphere we detect a cold spot at $(\theta, \phi) = (124°, 320°)$ and another one at $(\theta, \phi) = (124°, 78°)$. Both were already detected with the above mentioned directional and steerable wavelet as well as with a

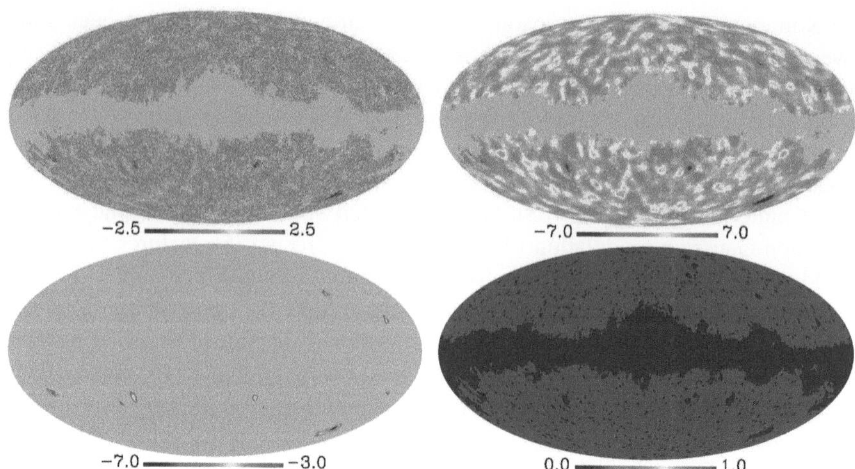

Fig. 4.8 The pixel-wise deviations $S_{i,r}$ of the primal (*upper left*) and of the smoothed scaling indices map (*upper right*), both based on the VW-band and the scale parameter $r = 0.2$. The plot in the *lower left* only shows the values ≤ -3.0 of the smoothed method. Except for the very small spots in the right part of this mapping, these regions are added to the KQ75-mask. The result is illustrated in the *lower right* plot

needlet analysis. The former one is a *hot* spot in these investigations. In our analysis, the latter spot actually appears as two spots close to each other, which is in agreement with [40]. We discover another southern cold spot at $(\theta, \phi) = (120°, 155°)$ which is very close to the mask. This spot represents a good example for the use of the mask filling method since it is situated at the edge of the non-masked region: The influence of the mask is diminishing the results of the calculation of the scaling indices in the area of this spot. This becomes obvious if one recalls the lower left plot of Fig. 4.1, in which the coordinates of the spot would be completely located in a "blue" region with low α-values. Since the results of the scaling indices of local features show a similar, namely lower-valued, behaviour, an overlapping like that could prevent the detection of such spots close to the mask. By using the mask filling method, the detection of this cold spot on the edge of the mask is equivalent to a detection in an unmasked region, and therefore reliable. The spot at $(\theta, \phi) = (136°, 173°)$, described by [48] and [40], is not traced in our analysis. In the northern hemisphere, our investigation shows two other cold spots at $(\theta, \phi) = (49°, 245°)$ and $(\theta, \phi) = (68°, 204°)$, which do not correspond with the so-called *northern cold spot* of [43], but with the results of [48], where again one of them is a hot spot. Also [40] locates one of these two spots. All these results were achieved with an analysis of the VW-band, but we find similar results in a single band analysis.

It is possible to define a new coordinate frame, including a new direction of the "north pole", such that all of these spots are contained in the "southern" hemisphere. This new north pole would then be located at $(\theta, \phi) = (51°, 21°)$.

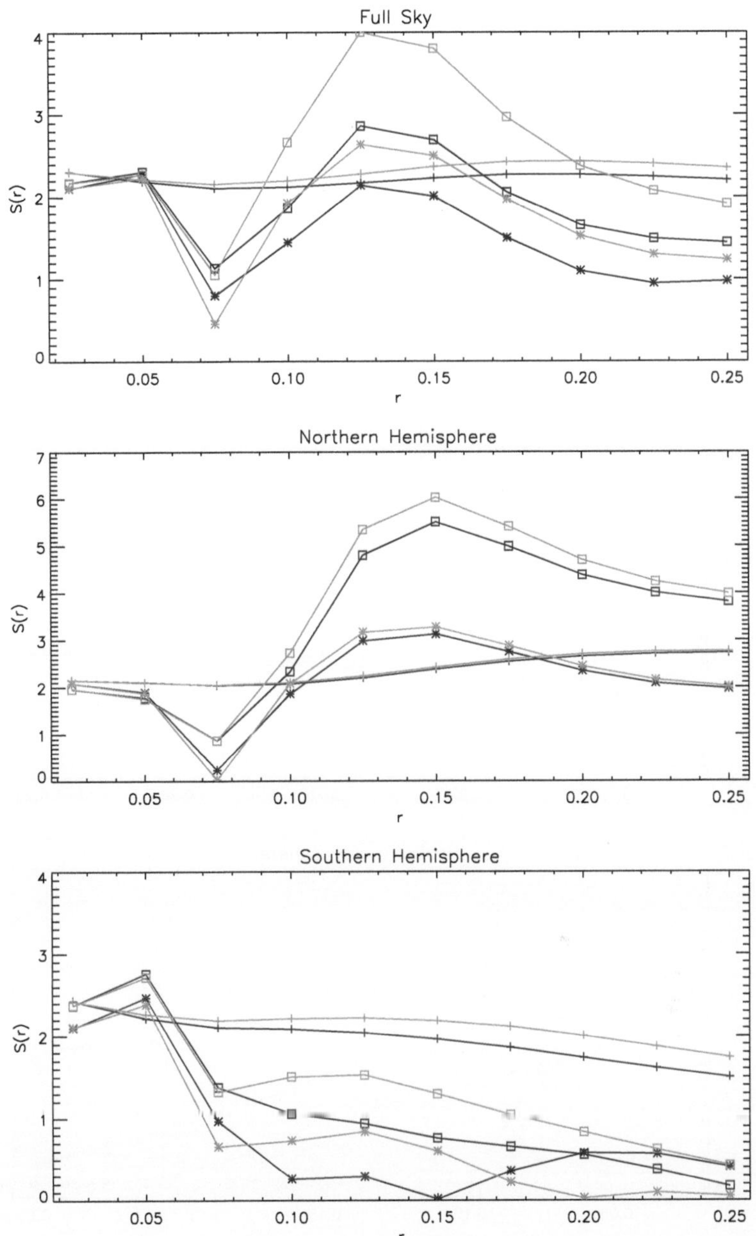

Fig. 4.9 The σ-normalised deviations of the mask-filling method for the original KQ75-mask (*blue*) and for the modified mask of the previous figure (*green*) in absolute values, plotted as a function of the scale parameter, whereby as above "+" denotes the mean, "$*$" the standard deviation and the boxes the χ^2-combination. The full sky as well as again the single hemispheres are considered. The *blue lines* exactly correspond to the *blue lines* of Fig. 4.5

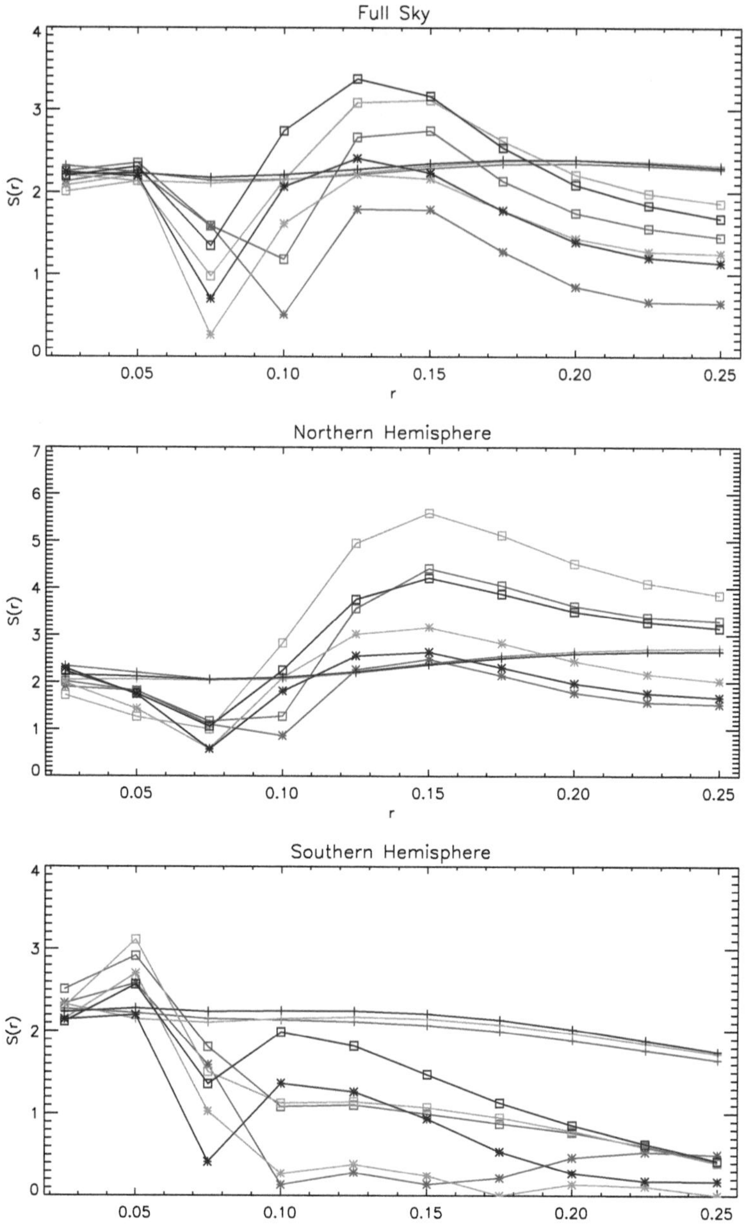

Fig. 4.10 Same as Fig. 4.9 but exclusive for the modified mask of Fig. 4.8 applied to the Q- (*red*), V- (*green*) and W-band (*blue*). This plot is associated with Fig. 4.6

If the considered spots really depend on some yet not completely understood, maybe secondary, physical effect, they should not be implemented in a testing for intrinsic non-Gaussianity. For this reason, we modify the KQ75-mask by additionally excluding all above mentioned spots. A small peculiarity at the edge of the mask next to the *cold spot* as well as three very small blurs in the right half of the lower left mollweide projection in Fig. 4.8 are not considered, since we regard their appearance as insufficient for being a distinctive feature. The modification of the KQ75-mask is illustrated in the lower right part of Fig. 4.8.

We now apply this new mask to the α-response of both the WMAP data as well as the simulations and repeat the analysis of Sect. 4.4.1. The results are illustrated in the Figs. 4.9 and 4.10 as well as in Table 4.4. A clear increase of $S(r)$ in comparison to the former analysis is evident. This heightening is in particular present in the southern hemisphere, where we detected more local features than in the north. The largest increase takes place in the co-added VW-band, where we now reach deviations of up to 4.0 for the χ^2-combination in a full-sky analysis (former maximum: 2.9) and to the extend of 6.0 in an analysis of the northern hemisphere (former maximum: 5.5). But also the single bands in Fig. 4.10 as well as all scale-independent diagonal χ^2-statistics in Table 4.4 show without exception a greater evidence for non-Gaussianity.

One could have expected to obtain higher values for $S(r)$ since the α-response of the WMAP data in comparison to the one of the simulations featured a shift to higher values (see Fig. 4.3): By now cutting out the local features, that exclusively consist of *cold* spots in terms of pixel-wise deviations, one excludes spots that showed lower values than the average of the simulations (see Eq. 4.10). Therefore, the shift to higher values becomes even larger, hence leading to a higher $S(r)$. Still, the exclusion of the spots is helpful and necessary, since these local anomalies could origin in some independent physical process, as mentioned above.

Table 4.4 Same as Table 4.2, but after excluding the cold spots via the modified KQ75-mask

	Full sky ($S/\%$)	Northern sky	Southern sky
$\chi^2_{\langle\alpha\rangle}$			
VW (mask-filled)	2.4/97.6	2.8/98.4	2.0/97.2
Q (mask-filled)	2.3/97.5	2.8/98.3	1.8/96.9
V (mask-filled)	2.3/97.6	2.7/98.1	1.9/97.1
W (mask-filled)	2.4/97.7	2.6/98.2	2.0/97.4
$\chi^2_{\sigma_\alpha}$			
VW (mask-filled)	2.6/96.7	4.8/99.8	0.2/97.2
Q (mask-filled)	1.3/90.6	2.5/97.0	0.6/82.1
V (mask-filled)	2.0/94.9	4.4/99.4	0.4/74.1
W (mask-filled)	2.2/96.2	3.1/98.0	0.4/78.9
$\chi^2_{\langle\alpha\rangle,\sigma_\alpha}$			
VW (mask-filled)	2.7/98.0	4.0/99.1	1.6/96.2
Q (mask-filled)	2.2/97.1	3.0/98.6	1.6/95.6
V (mask-filled)	2.4/97.3	3.7/98.9	1.5/95.3
W (mask-filled)	2.5/97.9	3.2/98.7	1.6/95.5

4.5 Summary

We performed a scaling index analysis of the WMAP 5-year data following up the investigations of [52]. For more realistic results around the mask, we additionally implemented a mask-filling method. By comparing the Q-, V-, W- and the co-added VW-band of the WMAP data with 1000 simulated maps per band, we (re)detected strong deviations from Gaussianity as well as asymmetries in the data, which can be summarized and interpreted as follows:

The scaling index values of the WMAP data are shifted to higher values and feature a higher variability than those of the simulations, especially in the northern hemisphere. This effect can be interpreted as less structure as well as more structural variations in the CMB signal compared to the corresponding Gaussian model. The results are confirmed by several statistics, that show deviations from Gaussianity of up to 5.9σ in the scale-dependent, and 5.5σ in the scale-independent case. These results are slightly lower applying the mask-filling method, and show high similarities within the different bands. In addition, we detected strong asymmetries by performing an analysis of rotated hemispheres: rotations pointing to northern directions show by far higher deviations from Gaussianity for the mean and the χ^2-analysis than rotations pointing to the south. Observing the standard deviation, we obtained a negative outcome in the north and a positive in the south. This implies that the north possesses a more consistent pattern than the simulations, while the south shows the converse behaviour. This feature is in line with later investigations of local features, where we detected more local anomalies in the southern than in the northern hemisphere.

Furthermore, we performed an analysis of local features by studying pixel-wise deviations from Gaussianity with and without a previous smoothing of the α-responses. For these investigations, we exclusively applied the mask filling method which can reduce the distorting effects on measurements like the scaling indices that appear when cutting out the masked regions. This mask-filling method eliminates the diluting effects on the border and therefore allows for an analysis of local features, which show a similar behaviour of lower outcome in the scaling index method. We detected the well-known *cold spot* and three additional spots in the southern as well as two spots in the northern hemisphere. Except for one single spot in the south, all findings are in agreement with former results of different investigations. Since these spots could origin on some yet not completely understood physical effect, we excluded them from the data set and repeated the former analysis. Instead of obtaining lower deviations, the results show an increase of non-Gaussianity in all bands. Therefore, the discovered local anomalies are not the reason of the global detection of non-Gaussianity, but were actually dampening the deviations on average. In former isotropic wavelet-based analyses, an exclusion of detected spots lessened the significance level of indications of non-Gaussianity [37]. Our new findings indicate in contrast, that the isotropic scaling index method can detect several different yet complementary aspects of the structural composition of the underlying data. The results of our investigation are in agreement with the steerable wavelet-based analysis in [79], where the non-Gaussianites were conserved after excluding the detected local anomalies.

4.6 Conclusions

The redetection of indications for non-Gaussianity of the WMAP 3-year data analysis leads to the conclusion that the observed results are not time-depending. In contrary, we can detect even higher deviations from the simulations which mimic the Gaussian properties of the best fit ΛCDM-model. Therefore, it is highly improbable for the results to be caused by effects related to short-term measurements.

In addition, the coherence between the different analysed bands implies that the foreground influence plays only a minor role but that the results are very unlikely to be truly of thermal origin.

Finally, the agreement of the detected spots with former investigations confirms the existence of these local anomalies.

The two most important tasks for future studies are: First, to identify possible reasons for the indications of non-Gaussianity, which could be possible with the attainment of more and more precise data, e.g. with the upcoming PLANCK-mission. Second, to figure out possible sources of the observed local features and thereby solving the question, if these anomalies are due to systematics or foreground effects or indeed represent variations in the CMB signal itself.

References

1. A.H. Guth, Phys. Rev. D **23**, 347 (1981)
2. A. Linde, Phys. Lett. **108B**, 389 (1982)
3. A. Albrecht, P. Steinhardt, Phys. Rev. Lett. **48**, 1220 (1982)
4. V. Acquaviva, N. Bartolo, S. Matarrese, A. Riotto, Nucl. Phys. B **667**, 119 (2003)
5. A.H. Linde, V. Mukhanov, Phys. Rev. D **56**, R535 (1997)
6. P.J.E. Peebles, ApJ Lett. **438**, L1 (1997)
7. F. Bernardeau, J.-P. Uzan, Phys. Rev. D **66**, 103506 (2002)
8. N. Bartolo, S. Matarrese, A. Riotto, Phys. Rev. D **65**, 103505 (2002)
9. D.H. Lyth, D. Wands, Phys. Lett. B **524**, 5 (2002)
10. D.H. Lyth, C. Ungarelli, D. Wands, Phys. Rev. D **67**, 023503 (2003)
11. J. Garriga, V. Mukhanov, Phys. Lett. B **458**, 219 (1999)
12. C. Armendariz-Picon, T. Damour, V. Mukhanov, Phys. Lett. B **458**, 209 (1999)
13. N. Bartolo, E. Komatsu, S. Matarrese, A. Riotto, Phys. Rep. **402**, 103 (2004)
14. M. LoVerde, A. Miller, S. Shandera, L. Verde, JCAP **4**, 14 (2008)
15. N. Kaiser, A. Stebbins, Nature **310**, 391 (1984)
16. N. Turok, D. Spergel, Phys. Rev. Lett. **64**, 2736 (1990)
17. N. Turok, ApJ Lett. **473**, L5 (1996)
18. R. Jeannerot, J. Rocher, M. Sakellariadou, Phys. Rev. D **68**, 103514 (2003)
19. E. Komatsu, N. Afshordi, N. Bartolo, D. Baumann, J.R. Bond, E.I. Buchbinder, C.T. Byrnes, X. Chen, D.J.H. Chung, A. Cooray, P. Creminelli, N. Dalal, O. Dore, R. Easther, A.V. Frolov,

K.M. Górski, M.G. Jackson, J. Khoury, W.H. Kinney, L. Kofman, K. Koyama, L. Leblond, J.-L. Lehners, J.E. Lidsey, M. Liguori, E.A. Lim, A. Linde, D.H. Lyth, J. Maldacena, S. Matarrese, L. McAllister, P. McDonald, S. Mukohyama, B. Ovrut, H.V. Peiris, C. Raeth, A. Riotto, Y. Rodriguez, M. Sasaki, R. Scoccimarro, D. Seery, E. Sefusatti, U. Seljak, L. Senatore, S. Shandera, E.P.S. Shellard, E. Silverstein, A. Slosar, K.M. Smith, A.A. Starobinsky, P.J. Steinhardt, F. Takahashi, M. Tegmark, A.J. Tolley, L. Verde, B.D. Wandelt, D. Wands, S. Weinberg, M. Wyman, A.P.S. Yadav, M. Zaldarriaga, Astro2010, The Astronomy and Astrophysics Decadal Survey. Science White Papers **158** (2009)

20. L.-Y. Chiang, P.D. Naselsky, O.V. Verkhodanov, M.J. Way, ApJ Lett. **590**, L65 (2003)
21. E. Komatsu, A. Kogut, M.R. Nolta, C.L. Bennett, M. Halpern, G. Hinshaw, N. Jarosik, M. Limon, S.S. Meyer, L. Page, D.N. Spergel, G.S. Tucker, L. Verde, E. Wollack, E.L. Wright, ApJS **148**, 119 (2003)
22. P. Coles, P. Dineen, J. Earl, D. Wright, MNRAS **350**, 989 (2004)
23. I.J. O'Dwyer, H.K. Eriksen, B.D. Wandelt, J.B. Jewell, D.L. Larson, K.M. Górski, A.J. Banday, S. Levin, P.B. Lilje, ApJ Lett. **617**, L99 (2004)
24. L.-Y. Chiang, P.D. Naselsky, P. Coles, ApJ **664**, 8 (2007)
25. H.K. Eriksen, G. Huey, A.J. Banday, K.M. Górski, J.B. Jewell, I.J. O'Dwyer, B.D. Wandelt, ApJ **665**, 1L (2007)
26. O. Rudjord, F.K. Hansen, X. Lan, M. Liguori, D. Marinucci, S. Matarrese, ApJ **701**, 369 (2009)
27. K.M. Smith, L. Senatore, M. Zaldarriaga, JCAP **0909**, 006 (2009)
28. H.K. Eriksen, F.K. Hansen, A.J. Banday, K.M. Górski, P.B. Lilje, ApJ **605**, 14 (2004)
29. F.K. Hansen, A.J. Banday, K.M. Górski, MNRAS **354**, 641 (2004)
30. F.K. Hansen, A.J. Banday, K.M. Górski, H.K. Eriksen, P.B. Lilije, ApJ **704**, 1448 (2009)
31. A. Bonaldi, S. Ricciardi, S. Leach, F. Stivoli, C. Baccigalupi, G. De Zotti, MNRAS **382**, 1791 (2007)
32. A. Bernui, Phys. Rev. D **78**, 063531 (2008)
33. A. Bernui, W.S. Hipolito-Ricaldi, MNRAS **389**, 1453 (2008)
34. J. Hoftuft, H.K. Eriksen, A.J. Banday, K.M. Górski, F.K. Hansen, P.B. Lilje, ApJ **699**, 985 (2009)
35. H.K. Eriksen, D.I. Novikov, P.B. Lilje, A.J. Banday, K.M. Górski, ApJ **612**, 64 (2004)
36. C.-G. Park, MNRAS **349**, 313 (2004)
37. P. Vielva, E. Martínes-González, R.B. Barreiro, J.L. Sanz, L. Cayón, ApJ **609**, 22 (2004)
38. F.K. Hansen, P. Cabella, D. Marinucci, N. Vittorio, ApJ Lett. **607**, L67 (2004)
39. K. Land, J. Magueijo, MNRAS **357**, 994 (2005)
40. D. Pietrobon, A. Amblard, A. Balbi, P. Cabella, A. Cooray, D. Marinucci, Phys. Rev. D **78**, 103504 (2008)
41. C. Räth, G. Morfill, G. Rossmanith, A.J. Banday, K.M. Gorski, PRL **102**, 131301 (2009)
42. V.G. Gurzadyan, A.A. Kocharyan, A&A Lett. **492**, L33 (2008)
43. V.G. Gurzadyan, A.E. Allahverdyan, T. Ghahramanyan, A.L. Kashin, H.G. Khachatryan, A.A. Kocharyan, H. Kuloghlian, S. Mirzoyan, E. Poghosian, G. Yegorian, A&A **497**, 343 (2009)
44. P. Mukherjee, Y. Wang, ApJ **613**, 51 (2004)
45. L. Cayón, L. Jin, A. Treaster, MNRAS **362**, 826 (2005)
46. M. Cruz, E. Martínes-González, P. Vielva, L. Cayón, MNRAS **356**, 29 (2005)
47. M. Cruz, L. Cayón, E. Martínes-González, P. Vielva, J. Jin, ApJ **655**, 11 (2007)
48. J.D. McEwen, M.P. Hobson, A.N. Lasenby, D.J. Mortlock, MNRAS **359**, 1583 (2005)
49. J.D. McEwen, M.P. Hobson, A.N. Lasenby, D.J. Mortlock, MNRAS Lett. **371**, L50 (2006)
50. J.D. McEwen, M.P. Hobson, A.N. Lasenby, D.J. Mortlock, MNRAS **388**, 659 (2008)
51. P. Vielva, Y. Wiaux, E. Martínes-González, P. Vandergheynst, MNRAS **381**, 932 (2007)
52. C. Räth, P. Schuecker, A.J. Banday, MNRAS **380**, 466 (2007)
53. G. Hinshaw, M.R. Nolta, C.L. Bennett, R. Bean, O. Dor, M.R. Greason, M. Halpern, R.S. Hill, N. Jarosik, A. Kogut, E. Komatsu, M. Limon, N. Odegard, S.S. Meyer, L. Page, H.V. Peiris, D.N. Spergel, G.S. Tucker, L. Verde, J.L. Weiland, E. Wollack, E.L. Wright, ApJS **170**, 288 (2007)

54. L. Page, G. Hinshaw, E. Komatsu, M.R. Nolta, D.N. Spergel, C.L. Bennett, C. Barnes, R. Bean, O. Dor, J. Dunkley, M. Halpern, R.S. Hill, N. Jarosik, A. Kogut, M. Limon, S.S. Meyer, N. Odegard, H.V. Peiris, G.S. Tucker, L. Verde, J.L. Weiland, E. Wollack, E.L. Wright, ApJS **170**, 335 (2007)
55. C.L. Bennett, M. Halpern, G. Hinshaw, N. Jarosik, A. Kogut, M. Limon, S.S. Meyer, L. Page, D.N. Spergel, G.S. Tucker, E. Wollack, E.L. Wright, C. Barnes, M.R. Greason, R.S. Hill, E. Komatsu, M.R. Nolta, N. Odegard, H.V. Peirs, L. Verde, J.L. Weiland, ApJS **148**, 1 (2003)
56. G. Hinshaw, J.L. Weiland, R.S. Hill, N. Odegard, D. Larson, C.L. Bennett, J. Dunkley, B. Gold, M.R. Greason, N. Jarosik, E. Komatsu, M.R. Nolta, L. Page, D.N. Spergel, E. Wollack, M. Halpern, A. Kogut, M. Limon, S.S. Meyer, G.S. Tucker, E.L. Wright, ApJS **180**, 225 (2009)
57. K.M. Górski, E. Hivon, A.J. Banday, B.D. Wandelt, F.K. Hansen, M. Reinecke, M. Bartelmann, ApJ **622**, 759 (2005)
58. B. Gold, C.L. Bennett, R.S. Hill, G. Hinshaw, N. Odegard, L. Page, D.N. Spergel, J.L. Weiland, J. Dunkley, M. Halpern, N. Jarosik, A. Kogut, E. Komatsu, D. Larson, S.S. Meyer, M.R. Nolta, E. Wollack, E.L. Wright, ApJ Suppl. Ser. **180**, 265 (2009)
59. C. Räth, W. Bunk, M.B. Huber, G.E. Morfill, J. Retzla, P. Schuecker, MNRAS **337**, 413 (2002)
60. C. Räth, P. Schuecker, MNRAS **344**, 115 (2003)
61. M. Gliozzi, W. Brinkmann, C. Räth, I.E. Papadakis, H. Negoro, H. Scheingraber, A&A **391**, 875 (2002)
62. M. Gliozzi, I.E. Papadakis, C. Räth, A&A **449**, 969 (2006)
63. F. Jamitzky, R.W. Stark, W. Bunk, S. Thalhammer, C. Räth, T. Aschenbrenner, G.E. Morfill, W.M. Heckl, Ultramicroscopy **86**, 241 (2001)
64. R. Monetti, H. Böhm, D. Müller, D. Newitt, S. Majumdar, E. Rummeny, T.M. Link, C. Räth, Proc. SPIE **5032**, 1777 (2003)
65. C. Räth, R. Monetti, J. Bauer, I. Sidorenko, D. Müller, M. Matsuura, E.-M. Lochmüller, P. Zysset, F. Eckstein, New J. Phys. **10**, 125010 (2008)
66. K. Tomita, Phys. Rev. D **72**, 10 (2005)
67. K. Tomita, K.T. Inoue, Phys. Rev. D **77**, 103522 (2008)
68. R.J. Adler, J.D. Bjorken, J.M. Overduin, gr-qc/0602102 (2006)
69. K.T. Inoue, J. Silk, ApJ **648**, 23 (2006)
70. K.T. Inoue, J. Silk, ApJ **664**, 650 (2007)
71. L. Rudnick, S. Brown, L. Williams, ApJ **671**, 40 (2007)
72. B.R. Granett, M.C. Neyrinck, I. Szapudi, ApJ **683**, L99 (2008)
73. M. Cruz, N. Turok, P. Vielva, E. Martínes-González, M. Hobson, Science **318**, 1612 (2007)
74. P.D. Naselsky, P.R. Christensen, P. Coles, O. Verkhodanov, D. Novikov, J. Kim, Astrophys. Bull. **65**, 101 (2010)
75. R.A. Battye, B. Garbrecht, A. Pilaftsis, JCAP **809**, 20 (2008)
76. J.A.R. Cembranos, A. de la Cruz-Dombriz, A. Dobado, A.L. Maroto, arXiv:0803.0694 (2008)
77. J. Garcia-Bellido, T. Haugbolle, JCAP **04**, 003 (2008)
78. I. Masina, A. Notari, JCAP **09**, 028 (2010)
79. Y. Wiaux, P. Vielva, R.B. Barreiro, E. Martínes-González, P. Vandergheynst, MNRAS **385**, 939 (2008)

Chapter 5
Surrogates and Scaling Indices Applied to the WMAP 7-Year Data

Inflationary models of the very early universe have proved to be in very good agreement with the observations of the linear correlations of the cosmic microwave background (CMB). While the simplest, single field, slow-roll inflation [1–3] predicts that the temperature fluctuations of the CMB correspond to a (nearly) Gaussian, homogeneous and isotropic random field, more complex models may give rise to non-Gaussianity [4–7]. Models in which the Lagrangian is a general function of the inflaton and powers of its first derivative [8, 9] can lead to scale-dependent non-Gaussianities, if the sound speed varies during inflation. Similarily, string theory models that give rise to large non-Gaussianity have a natural scale dependence [10]. If the scale dependence of non-Gaussian signatures plays an important role in theory, the conventional (global) parametrisation of non-Gaussianity via f_{NL} is no longer sufficient to describe the level of non-Gaussianity and to discriminate between different models. f_{NL} must at least become scale dependent—if this parametrisation is sufficient at all. But first of all such scale-dependent signatures have to be identified.

Possible deviations from Gaussianity have been investigated in studies based on e.g. the WMAP data of the CMB (see [11] and references therein) and claims for the detection of non-Gaussianities and other anomalies (see e.g. [12–20]) have been made. These studies have in common that the level of non-Gaussianity is assessed by comparing the results for the measured data with a set of simulated CMB-maps which were generated on the basis of the standard cosmological model and/or specific assumptions about the nature of the non-Gaussianities

On the other hand, it is possible to develop model-independent tests for higher order correlations (HOCs) by applying the ideas of constrained randomisation [21–23], which have been developed in the field of nonlinear time series analysis [24]. The basic formalism is to compute statistics sensitive to HOCs for the original data set and for an ensemble of surrogate data sets, which mimic the linear properties of the

---

Original publication: C. Räth, G. E. Morfill, G. Rossmanith, A. J. Banday, K. M. Górski, *A model-independent test for scale-dependent non-Gaussianities in the cosmic microwave background*, PRL, **102**, 131301 (2009).

G. Rossmanith, *Non-linear Data Analysis on the Sphere*, Springer Theses, DOI: 10.1007/978-3-319-00309-2_5, © Springer International Publishing Switzerland 2013

original data. If the computed measure for the original data is significantly different from the values obtained for the set of surrogates, one can infer that the data contain HOCs.

Based on these ideas we present in this *Letter* a new method for generating surrogates allowing for probing scale-dependent non-Gaussianities.

Our study is based on the WMAP data of the CMB. Since our method in its present form requires full sky coverage to ensure the orthogonality of the set of basis functions Y_{lm} we used the five-year "foreground-cleaned" Internal Linear Combination (ILC) map (WMAP5) [25] generated and provided[1] by the WMAP-team. For comparison we also included the maps produced by Tegmark et al. [26, 27], namely the three year cleaned map (TOHc3) and the Wiener-filtered cleaned map (TOHw3),[2] which were generated pursuing a different approach for foreground cleaning. Since the Gaussianity of the temperature distribution and the randomness of the set of Fourier phases are a necessary prerequisite for the application of our method we performed the following preprocessing steps. First, the maps were remapped onto a Gaussian distribution in a rank-ordered way. By applying this remapping we automatically focus on HOCs induced by the spatial correlations in the data while excluding any effects coming from deviations of the temperature distribution from a Gaussian one. To ensure the randomness of the set of Fourier phases we performed a rank-ordered remapping of the phases onto a set of uniformly distributed ones followed by an inverse Fourier transformation. These two preprocessing steps result in minimal changes to the ILC map (the maps remain highly correlated with cross-correlations $c > 0.95$). The main effect is the removal of significant outliers in the temperature distribution.

To test for scale-dependsent non-Gaussianities in a model-independent way we propose the following two-step procedure. Without loss of generality we restrict the description of the method and all subsequent analyses to the case of non-Gaussianities on large scales. Consider a CMB map $T(\theta, \phi)$, where $T(\theta, \phi)$ is Gaussian distributed and calculate its Fourier transform. The complex valued Fourier coefficients a_{lm}, $a_{lm} = \int d\Omega_n T(n) Y^*_{lm}(n)$ can be written as $a_{lm} = |a_{lm}| e^{i\phi_{lm}}$ with $\phi_{lm} = \arctan(Im(a_{lm})/Re(a_{lm}))$. The linear or Gaussian properties of the underlying random field are contained in the absolute values $|a_{lm}|$, whereas all HOCs—if present—are encoded in the phases ϕ_{lm} and the correlations among them. First, we generate a first order surrogate map, in which any phase correlations for the scales, which are not of interest (here: the small scales), are randomised. This is achieved by a random shuffle of the phases ϕ_{lm} for $l > l_{cut}, 0 < m \leq l$, where $l_{cut} = 10, 15, 20, 25, 30$ in this *Letter* and by performing an inverse Fourier transformation (Fig. 5.1). Second, N ($N = 500$ for $l_{cut} = 20$, $N = 100$ otherwise) realisations of second order surrogate maps are generated for the first order surrogate map, in which the remaining phases ϕ_{lm} with $1 < l \leq l_{cut}, 0 < m \leq l$ are shuffled while the already randomised phases for the small scales are preserved.

---

[1] http://lambda.gsfc.nasa.gov/

[2] http://space.mit.edu/home/tegmark/wmap.html

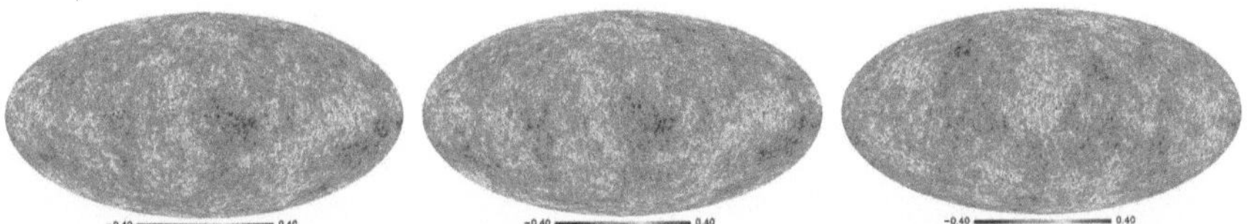

Fig. 5.1 ILC map after remapping of the temperatures and phases (*left*). First order (*middle*) and respective second order surrogate (*right*) for $l_{cut} = 20$. Note the resemblance of the first order surrogate with the ILC map at large scales

Figure 5.1 shows a realisation of a second order surrogate map after inverse Fourier transformation. Note that the Gaussian properties of the *remapped* ILC map, which are given by $|a_{lm}|$, are *exactly* preserved in all surrogate maps. Finally, for calculating higher order statistics the maps were degraded to $N_{side} = 256$ and residual monopole and dipole contributions were subtracted. To compare the two classes of surrogates, we calculate local statistics in the spatial domain, namely scaling indices (SIM) as described in Räth et al. [18]. In brief, scaling indices estimate local scaling properties of a point set P. The spherical CMB data can be represented as a three-dimensional point distribution $P = \vec{p}_i = (x_i, y_i, z_i)$, $i = 1, \ldots, N_{pixels}$ by transforming the temperature fluctuations into a radial jitter. For each point $\vec{p}_i$ the local weighted cumulative point distribution ρ is calculated $\rho(\vec{p}_i, r) = \sum_{j=1}^{N_{pixels}} e^{-(\frac{d_{ij}}{r})^2}$, $d_{ij} = \|\vec{p}_i - \vec{p}_j\|$. The weighted scaling indices $\alpha(\vec{p}_i, r)$ are then obtained by calculating the logarithmic derivative of $\rho(\vec{p}_i, r)$ with respect to r, $\alpha(\vec{p}_i, r) = \frac{\partial \log \rho(\vec{p}_i, r)}{\partial \log r}$. For each pixel we calculated scaling indices for ten different scales, $r_1 = 0.025, \ldots, r_{10} = 0.25$ in the notation of [18]. For each scale we calculate the mean ($\langle \alpha \rangle$) and standard deviation (σ_α) of the scaling indices $\alpha(\vec{p}_i, r)$ derived from a set of pixels belonging to rotated hemispheres or the full sky. To investigate the correlations between the scaling indices and temperature fluctuations, we also considered the standard deviation (σ_T) for the mere temperature distribution of the respective sky regions.

The differences of the two classes of surrogates are quantified by the σ-normalised deviation

$$S(Y) = (Y_{surro1} - \langle Y_{surro2} \rangle)/\sigma_{Y_{surro2}},$$

$Y = \sigma_T, \langle \alpha \rangle, \sigma_\alpha, \chi^2$ (surro1: first order surrogate, surro2: second order surrogate) and the significance levels $SL = 1 - p$, where p is the fraction of second order surrogates, which have a higher (lower) Y than the first order surrogate. χ^2 denotes diagonal χ^2-statistics, which we obtain by combining $\langle \alpha \rangle$, σ_α for a given scale r_i, i.e.

$$\chi^2(r_i) = \sum_{j=1}^{2} \left[\frac{X_j(r_i) - \langle X_j(r_i) \rangle}{\sigma_{X_j(r_i)}} \right]^2,$$

with $X_1 = \langle \alpha \rangle$, $X_2 = \sigma_\alpha$ and $\langle X_j \rangle$, σ_{X_j} derived from the N realisations of second order surrogates. As scale-independent measure we also consider χ^2 as obtained by

Fig. 5.2 Deviation S as derived from rotated upper hemispheres for σ_T (*left*) and $\langle\alpha(r_{10})\rangle$ (*right*) for the WMAP5 map and $l_{cut} = 20$. The z-axis of the respective rotated reference frame pierces the centre of the respective colour-coded pixel. 768 rotated hemispheres, which correspond to number of coloured pixels, were considered. (For a more detailed description of this visualisation technique see e.g. [14, 18])

summing over the scales ($N_r = 10$),

$$\chi^2 = \sum_{i=1}^{N_r} \sum_{j=j_1}^{j_2} \left[\frac{X_j(r_i) - \langle X_j(r_i)\rangle}{\sigma_{X_j(r_i)}} \right]^2 ,$$

for one single measure ($j_1 = 1, j_2 = 1; j_1 = 2, j_2 = 2$) and the two measures ($j_1 = 1, j_2 = 2$). Figure 5.2 shows $S(\sigma_T)$ and $S(\langle\alpha(r_{10})\rangle)$ derived from pixels belonging to the respective upper hemispheres for 768 rotated reference frames. Statistically significant signatures for non-Gaussianity and ecliptic hemispherical asymmetries become immediately obvious, whereby these signatures can solely be induced by large scale HOCs. Although $S(\sigma_T)$ and $S(\langle\alpha(r_{10})\rangle)$ are spatially highly (anti-)correlated ($c = -0.95$), the two effects are nevertheless complementary to each other in the sense that a systematically lower/higher σ_T would lead to a lower/higher $\langle\alpha(r_{10})\rangle$ and not to the observed higher/lower value for the first order surrogate map. These systematically shifted scaling indices are a generic feature present in all three maps (Fig. 5.3). Although the probability densities $P(\alpha(r_{10}))$ are different due to the smoothing or Wiener-filtering for the three maps, the shifts of the first order surrogate relative to its second order surrogates can be found in all three cases. We also cross-correlated the deviation maps shown in Fig. 5.2 derived from the three input maps and always obtained $c \geq 0.98$ for the correlation coefficient. These systematic deviations lead to significant detections of non-Gaussianities which are shown in Fig. 5.4 and summarised for $l_{cut} = 20$ in Tables 5.1 and 5.2. The most significant and most stable results are found for $\langle\alpha\rangle$ at larger radii, where for all three maps none of the 500 second order surrogates had a higher (upper hemisphere) or lower (lower hemisphere) value than the respective first order surrogate, leading to a significance level $SL > 99.8\,\%$ for $\langle\alpha(r_{10})\rangle$. Also the combined measure $\chi^2_{\langle\alpha\rangle}$ yields deviations S ranging from 5.2 up to 7.9, which represent one of the most significant detection of non-Gaussianity in the WMAP data to date. We estimated how varying l_{cut} values affect the results and found that both the non-Gaussianities and asymmetries are detected for all considered l_{cut}, where the highest deviations

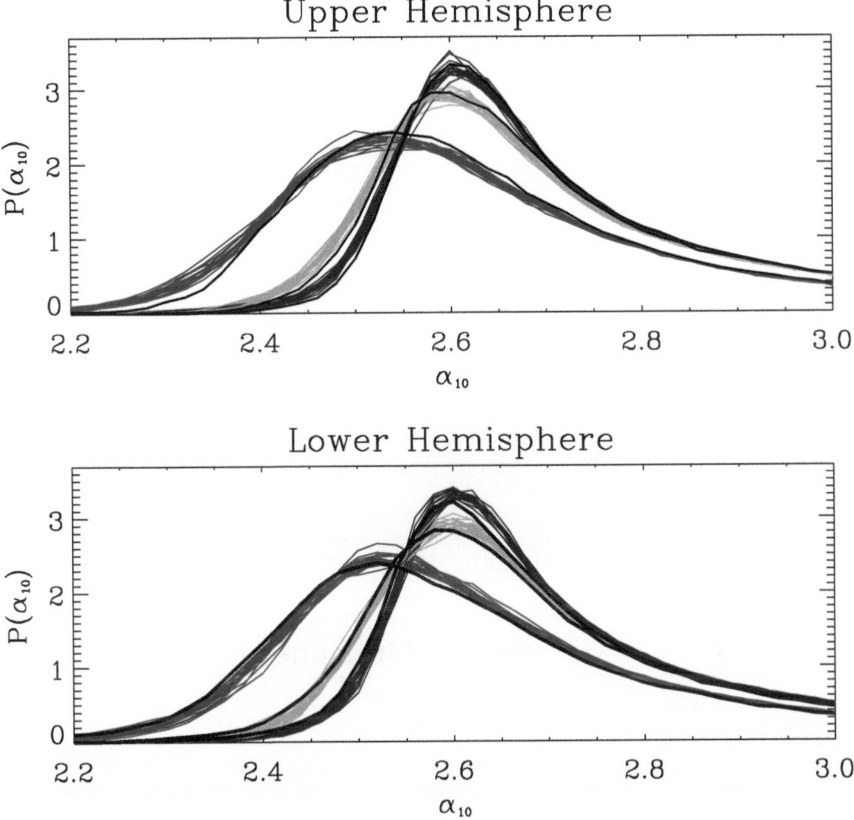

Fig. 5.3 Probability density $P(\alpha(r_{10}))$ for the surrogates of the WMAP5 (*blue*), TOHw3 (*yellow*) and TOHc3 (*red*) map for the rotated upper and lower hemisphere and $l_{cut} = 20$. The *black lines* denote the respective first order surrogate. The reference frame is chosen such that the difference $\Delta S = S_{up} - S_{low}$ between the upper and lower hemisphere becomes maximal for $\langle \alpha(r_{10}) \rangle$ regarding the WMAP5 surrogates

are obtained for $l_{cut} = 20$. Although S becomes considerably smaller for $l_{cut} = 10$, we can still detect the non-Gaussianities with $SL > 99.0\,\%$, which is larger than the results reported in [28] ($SL = 95\,\%$), where also $l_{cut} = 10$ was used. We performed the same analyses for the coadded WMAP foreground template maps and for simulations using the best fit ΛCDM power spectrum and WMAP-like noise and beam properties. We found in none of these cases significant signatures as reported above. Details about these studies are deferred to a longer forthcoming publication.

In conclusion, we demonstrated the feasibility to generate new classes of surrogate data sets preserving the power spectrum and partly the information contained in the Fourier phases, while all other HOCs are randomised. We found significant evidence for both asymmetries and non-Gaussianities on large scales in the WMAP data of the CMB using scaling indices as test statistics. The novel statistical test involving new

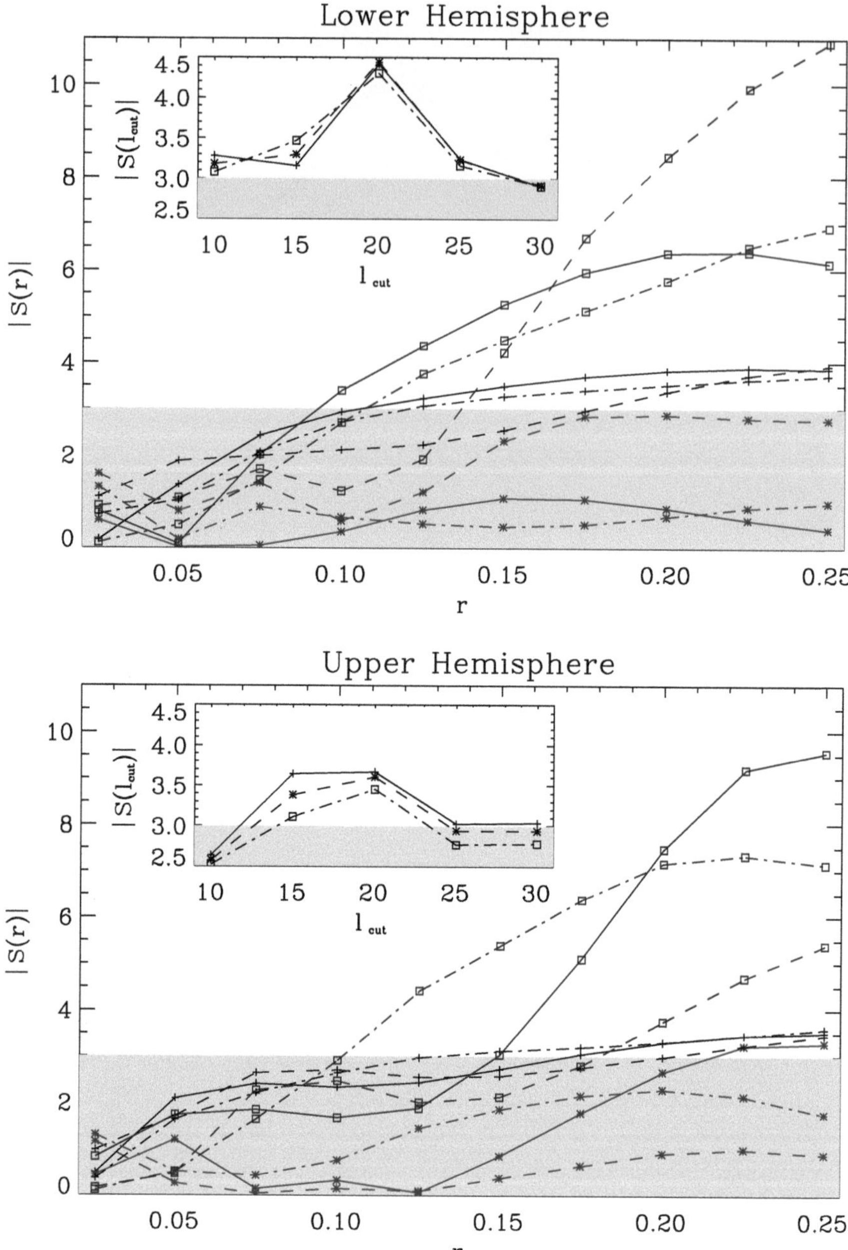

Fig. 5.4 Deviations $|S(r)|$ for the rotated upper and lower hemisphere for $\langle\alpha\rangle$ (*black*), σ_α (*blue*) and a χ^2-combination of $\langle\alpha\rangle$ and σ_α (*red*) ($l_{cut} = 20$, $N = 500$). The solid (*dashed, dashed-dotted*) *lines* denote the WMAP5 (TOHw3, TOHc3) map. The *shaded region* indicates the 3σ significance interval. The insets show the results for $\langle\alpha(r_{10})\rangle$, $\langle\alpha(r_9)\rangle$ and $\langle\alpha(r_8)\rangle$ (*solid, dashed, dashed-dotted*) as a function of l_{cut} for the WMAP5 map (here: $N = 100$)

Table 5.1 S/SL upper hemisphere

	WMAP5 (S/SL)	TOHc3 (S/SL)	TOHw3 (S/SL)
σ_T	$-2.8/99.8$	$-3.0/>99.8$	$-2.9/>99.8$
$\langle \alpha(r_{10}) \rangle$	$3.5 / >99.8$	$3.5 / >99.8$	$3.6 / >99.8$
$\chi^2_{\langle \alpha \rangle}$	$5.7 / 99.8$	$5.2 / 99.6$	$7.0 / >99.8$
$\chi^2_{\sigma_\alpha}$	$3.1 / 99.2$	$-0.7/ 74.4$	$2.1 / 95.4$
$\chi^2_{\langle \alpha \rangle, \sigma_\alpha}$	$6.1 / > 99.8$	$3.6 / 99.0$	$6.4 / > 99.8$

Table 5.2 S/SL lower hemisphere

	WMAP5 (S/SL)	TOHc3 (S/SL)	TOHw3 (S/SL)
σ_T	$2.7/99.8$	$2.9/>99.8$	$2.8/99.8$
$\langle \alpha(r_{10}) \rangle$	$-3.9/ > 99.8$	$-3.9/ > 99.8$	$-3.7/ > 99.8$
$\chi^2_{\langle \alpha \rangle}$	$7.9 />99.8$	$5.4/99.8$	$7.3/>99.8$
$\chi^2_{\sigma_\alpha}$	$-0.7/76.4$	$4.4 / 99.6$	$-0.6/67.0$
$\chi^2_{\langle \alpha \rangle, \sigma_\alpha}$	$5.8 / 99.8$	$6.3/>99.8$	$5.2/>99.8$

classes of surrogates allows for an unambigous relation of the signatures identified in real space with scale-dependent HOCs, which are encoded in the respective Fourier phase correlations. Our results, which are consistent with previous findings [12–16, 18, 28] but also extend to smaller scales than those reported in [17] ($l_{cut} = 3$), [28] ($l_{cut} = 10$) and [20] ($l_{cut} \leq 3$), point towards a violation of statistical isotropy and Gaussianity. Such features would disfavour canonical single-field slow-roll inflation—unless there is some undiscovered systematic error in the collection or reduction of the CMB data or yet unknown foreground contributions. Thus, at this stage it is too early to claim the detected HOCs as cosmological and further tests are required to elucidate the true origin of the detected anomalies. Their existence in the three maps might, however, be suggestive.

In either case the proposed statistical method offers an efficient tool to develop model-independent tests for scale-dependent non-Gaussianities. Due to the generality of this technique it can be applied to any signal, for which the analysis of scale-dependent HOCs is of interest.

Many of the results in this paper have been obtained using HEALPix [29]. We acknowledge the use of LAMBDA. Support for LAMBDA is provided by the NASA Office of Space Science.

References

1. A.H. Guth, Phys. Rev. D **23**, 347 (1981)
2. A. Linde, Phys. Lett. **108B**, 389 (1982)
3. A. Albrecht, P. Steinhardt, Phys. Rev. Lett. **48**, 1220 (1982)
4. A.H. Linde, V. Mukhanov, Phys. Rev. D **56**, R535 (1997)

5. P.J.E. Peebles, ApJ Lett. **438**, L1 (1997)
6. F. Bernardeau, J.-P. Uzan, Phys. Rev. D **66**, 103506 (2002)
7. V. Acquaviva, N. Bartolo, S. Matarrese, A. Riotto, Nucl. Phys. B **667**, 119 (2003)
8. C. Armendariz-Picon, T. Damour, V. Mukhanov, Phys. Lett. B **458**, 209 (1999)
9. J. Garriga, V. Mukhanov, Phys. Lett. B **458**, 219 (1999)
10. M. LoVerde, A. Miller, S. Shandera, L. Verde, JCAP **4**, 14 (2008)
11. E. Komatsu, J. Dunkley, M.R. Nolta, C.L. Bennett, B. Gold, G. Hinshaw, N. Jarosik, D. Larson, M. Limon, L. Page, D.N. Spergel, M. Halpern, R.S. Hill, A. Kogut, S.S. Meyer, G.S. Tucker, J.L. Weiland, E. Wollack, E.L. Wright, ApJ Suppl. Ser. **180**, 330 (2009)
12. C.-G. Park, MNRAS **349**, 313 (2004)
13. H.K. Eriksen, D.I. Novikov, P.B. Lilje, A.J. Banday, K.M. Górski, ApJ **612**, 64 (2004)
14. F.K. Hansen, A.J. Banday, K.M. Górski, MNRAS **354**, 641 (2004)
15. H.K. Eriksen, A.J. Banday, K.M. Gorski, P.B. Lilje, ApJ **622**, 58 (2005)
16. H.K. Eriksen, A.J. Banday, K.M. Gorski, F.K. Hansen, P.B. Lilje, ApJ Lett. **660**, L81 (2007)
17. A. Oliveira-Costa, M. Tegmark, M. Zaldarriaga, A. Hamilton, Phys. Rev. D **69**, 063516 (2004)
18. C. Räth, P. Schuecker, A.J. Banday, MNRAS **380**, 466 (2007)
19. J.D. McEwen, M.P. Hobson, A.N. Lasenby, D.J. Mortlock, MNRAS **388**, 659 (2008)
20. C.J. Copi, D. Huterer, D.J. Schwarz, G.D. Starkman, MNRAS **399**, 295 (2009)
21. M.P. Pompilio, F.R. Bouchet, G. Murante, A. Provenzale, ApJ **449**, 1 (1995)
22. C. Räth, W. Bunk, M.B. Huber, G.E. Morfill, J. Retzla, P. Schuecker, MNRAS **337**, 413 (2002)
23. C. Räth, P. Schuecker, MNRAS, **344**, 115 (2003)
24. J. Theiler, S. Eubank, A. Longtin, B. Galdrikian, J.D. Farmer, Physica D **58**, 77 (1992)
25. B. Gold, C.L. Bennett, R.S. Hill, G. Hinshaw, N. Odegard, L. Page, D.N. Spergel, J.L. Weiland, J. Dunkley, M. Halpern, N. Jarosik, A. Kogut, E. Komatsu, D. Larson, S.S. Meyer, M.R. Nolta, E. Wollack, E.L. Wright, ApJ. Suppl. Ser. **180**, 265 (2009)
26. M. Tegmark, A. Oliveira-Costa, A. Hamilton, Phys. Rev. D **68**, 123523 (2003)
27. A. de Oliveira-Costa, M. Tegmark, Phys. Rev. D **74**, 023005 (2006)
28. L.-Y. Chiang, P.D. Naselsky, P. Coles, ApJ **664**, 8 (2007)
29. K.M. Górski, E. Hivon, A.J. Banday, B.D. Wandelt, F.K. Hansen, M. Reinecke, M. Bartelmann, ApJ **622**, 759 (2005)

Chapter 6
Extending the Analysis of the WMAP 7-Year Data

6.1 Introduction

The Cosmic Microwave Background (CMB) radiation represents the oldest observable signal in the Universe. Since this relic radiation has its origin just 380000 years after the Big Bang when the CMB photons were last scattered off electrons, this radiation is one of the most important sources of information to gain more knowledge about the very early Universe. Estimating the linear correlations of the temperature fluctuations in the CMB as measured e.g. with the WMAP satellite by means of the power spectrum has yielded very precise determinations of the parameters of the standard ΛCDM cosmological model like the age, the geometry and the matter and energy content of the Universe [1, 2].

Analyzing CMB maps by means of the power spectrum represents an enormous compression of information contained in the data from approx. 10^6 temperature values to roughly 1000 numbers for the power spectrum. It has often been pointed out ([3] and references therein) that this data compression is lossless and thus fully justified, if and *only if* the statistical distribution of the observed fluctuations is a Gaussian distribution with random phases. Any information that is contained in the phases and the correlations among them, is not encoded in the power spectrum, but has to be extracted from measurements of higher-order correlation (HOC). Thus, the presence of phase correlations may be considered as an unambiguous evidence of non-Gaussianity (NG). Otherwise, non-Gaussianity can only be defined by the negation of Gaussianity.

Primordial NG represents one way to test theories of inflation with the ultimate goal to constrain the shape of the potential of the inflaton field(s) and their possible (self-)interactions. While the simplest single field slow roll inflationary scenario predicts that fluctuations are nearly Gaussian [4–6], a variety of more complex models

Original publication: C. Räth, A. J. Banday, G. Rossmanith, H. Modest, R. Sütterlin, K. M. Górski, J. Delabrouille and G. E. Morfill, *Scale-dependent non-Gaussianities in the WMAP data as identified by using surrogates and scaling indices*, MNRAS, **415**, 2205 (2011).

G. Rossmanith, *Non-linear Data Analysis on the Sphere*, Springer Theses, DOI: 10.1007/978-3-319-00309-2_6, © Springer International Publishing Switzerland 2013

predict deviations from Gaussianity [7–10]. Models in which the Lagrangian is a general function of the inflaton and powers of its first derivative [11, 12] can lead to scale-dependent non-Gaussianities, if the sound speed varies during inflation. Similarly, string theory models that give rise to large non-Gaussianity have a natural scale dependence [13, 14]. Also, NGs put strong constraints on alternatives to the inflationary paradigm [15, 16].

Given the plethora of conceivable scenarios for the very early Universe, it is worth first checking what is in the data in a model-independent way. Further, such a model-independent approach has a large discovery potential to detect yet unexpected fingerprints of nonlinear physics in the early universe. Thus, a detection of possibly scale-dependent non-Gaussianity being encrypted in the phase correlations in the WMAP data would be of great interest. While a detection of non-Gaussianity could be indicative of an experimental systematic effect or of residual foregrounds, it could also point to new cosmological physics.

The investigations of deviations from Gaussianity in the CMB (see [1] and references therein) and claims for the detection of non-Gaussianitiy and a variety of other anomalies like hemispherical asymmetries, lack of power at large angular scales, alignment of multipoles, detection of the Cold Spot etc. (see e.g. [17–30]) have been made, where the statistical significance of some of the detected signatures is still subject to discussion [31, 32]. These studies have in common that the level of non-Gaussianity is assessed by comparing the results for the measured data with simulated CMB-maps which were generated on the basis of the standard cosmological model and/or specific assumptions about the nature of the non-Gaussianities as parametrised with e.g. the scalar, scale-independent parameter f_{nl}. Other studies focused on the detection of signatures in the distribution of Fourier phases [33–36] representing deviations from the random phase hypothesis for Gaussian random fields. These model-independent tests also revealed signatures of NGs. Pursuing this approach one can go one step further and investigate possible phase correlations and their relation to the morphology of the CMB maps by means of so-called surrogate maps.

This technique of surrogate data sets [37] was originally developed for nonlinear time series analysis. In this field of research complex systems like the climate, stock-market, heart-beat variability, etc. are analyzed (see e.g. [38] and references therein). For those systems a full modelling is barely or not possible. Therefore, statistical methods of constrained randomization involving surrogate data sets were developed to infer some information about the nature of the underlying physical process in a completely data-driven, i.e. model-independent way. One of the first and most basic question here is whether a (quasiperiodic) process is completely linear or whether also weak nonlinearities can be detected in the data. The basic formalism to answer this question is to compute statistics sensitive to HOCs for the original data set and for an ensemble of surrogate data sets, which mimic the linear properties of the original time series while wiping out all phase correlations. If the computed measure for the original data is significantly different from the values obtained for the set of surrogates, one can infer that the data contain HOCs.

Extensions of this formalism to three-dimensional galaxy distributions [39] and two-dimensional simulated flat CMB maps [40] have been proposed and discussed.

By introducing a more sophisticated two-step surrogatization scheme for full-sky CMB observations it has become possible to also test for scale-dependent NG in a model-independent way [41]. Probing NG on the largest scales ($l < 20$) yielded highly significant signatures for both NG and ecliptic hemispherical asymmetries.

In this chapter, we apply the method of constrained randomization to the WMAP five year and seven year data in order to test for scale-independent and scale-dependent non-Gaussianity up to $l = 300$ as encoded in the Fourier phase correlations. Further, this work fully recognises the need to rule out foregrounds and systematic artefacts as the origin of the detections (as advised by [32]). Therefore, a large part of our analyses is dedicated to various checks on systematics to single out possible causes of the detected anomalies.

The chapter is organized as follows: In Sect. 6.2 we briefly describe the observational and simulated data we use in our study. The method of constrained randomization is reviewed in some detail in Sect. 6.3. Scaling indices, which we use as test statistic, and the statistics derived out of them are discussed in Sect. 6.4. In Sect. 6.5 we present our results and we draw our conclusions in Sect. 6.6.

6.2 Data Sets

We used the seven years foreground-cleaned internal linear combination (ILC) map [42] generated and provided by the WMAP team[1] (in the following: ILC7). For comparison we also included the map produced by [43], namely the five years needlet-based ILC map, which has been shown to be significantly less contaminated by foreground and noise than other existing maps obtained from WMAP data (in the following: NILC5).

To check for systematics we also analyzed the following set of maps:

(1) *Uncorrected ILC map*

The ILC map is a weighted linear combination of the 5 frequency channels that recovers the CMB signal. The weights are derived by requiring minimum variance in a given region of the sky under the constraint that the sum of the weights is unity. Such weights, however, cannot null an arbitrary foreground signal with a non-blackbody frequency spectrum, thus some residuals due to Galactic emission will remain. The WMAP team attempts to correct for this "bias" with an estimation of the residual signal based on simulations and a model of the foreground sky. Our uncorrected map (UILC7 in the following) is simply the ILC without applying this correction, computed from the weights provided in [42] and the 1-degree smoothed WMAP data.

(2) *Asymmetric beam map*

Beam asymmetries may result in statistically anisotropic CMB maps. To asses these effects on the signatures of scale-dependent NGs and their (an-)isotropies

---

[1] http://lambda.gsfc.nasa.gov

we make use of the publicly available CMB sky simulations including the effects of asymmetric beams [44]. Specifically we analyse a simulated map of the V1-band, because this band is considered to have the least foreground contamination.

(3) *Simulated coadded VW-band map*

To make sure that neither systematic effects are induced by the method of constrained randomization nor the WMAP-like beam and noise properties lead to systematic deviations from Gaussianity we include in our analysis a co-added VW-map as obtained using the standard ΛCDM best fit power spectrum and WMAP-like beam and noise properties. Note that this map did not undergo the ILC-map making procedure.

(4) *Simulated ILC map*

Simulated sky maps result from processing a simulated differential time-ordered data (TOD) stream through the same calibration and analysis pipeline that is used for the flight data. The TOD is generated by sampling a reference sky that includes both CMB and Galactic foregrounds with the actual flight pointings, and adding various instrumental artefacts. We have then processed the individual resulting data into 7 separate simulated yearly ILC maps, plus a 7-year merge. It is worth noting that, if the yearly frequency-averaged maps are combined into ILCs using the [42] 7-year weights per region, then the resulting ILCs show clear Galactic plane residuals. This reflects the fact that the simulated data has a different CMB realisation to the observed sky, and may additionally represent a mismatch between the simulated foreground properties and the true sky in the Galactic plane. Instead, we analyse the 7-year merged simulated data to compute the ILC weights for the simulations, then apply to all yearly data sets separately. However, the derived weights are quite different from the WMAP7 ones, which would imply different noise properties in the simulated ILC data compared to the real data. Care should be exercised for any results that are sensitive to the specific noise pattern.

(5) *Difference ILC map*

Finally, we consider the difference map (year 7–year 6) from yearly ILC-maps computed using the same weights and regions as the 7-year data set from [42]. No debiasing has been applied. With this map we estimate what effect possible ILC-residuals may have on the detection of NGs.

6.3 Generating Surrogate Maps

To test for scale-dependent non-Gaussianities in a model-independent way we apply a two-step procedure that has been proposed and discussed in [41]. Let us describe the various steps for generating surrogate maps in more detail:

Consider a CMB map $T(\theta, \phi)$, where $T(\theta, \phi)$ is Gaussian distributed and its Fourier transform. The Fourier coefficients a_{lm} can be written as $a_{lm} = |a_{lm}|e^{i\phi_{lm}}$ with $\phi_{lm} = \arctan(\mathrm{Im}(a_{lm})/\mathrm{Re}(a_{lm}))$. The linear or Gaussian properties of the underlying random field are contained in the absolute values $|a_{lm}|$, whereas all

HOCs—if present—are encoded in the phases ϕ_{lm} and the correlations among them. Having this in mind, a versatile approach for testing for scale dependent non-Gaussianities relies on a scale-dependent shuffling procedure of the phase correlations followed by a statistical comparison of the so-generated surrogate maps.

However, the Gaussianity of the temperature distribution and the randomness of the set of Fourier phases in the sense that they are uniformly distributed in the interval $[-\pi, \pi]$, are a necessary prerequisite for the application of the surrogate-generating algorithm, which we propose in the following. To fulfil these two conditions, we perform the following preprocessing steps. First, the maps are remapped onto a Gaussian distribution in a rank-ordered way. This means that the amplitude distribution of the original temperature map in real space is replaced by a Gaussian distribution in a way that the rank-ordering is preserved, i.e. the lowest value of the original distribution is replaced with the lowest value of the Gaussian distribution etc. By applying this remapping we automatically focus on HOCs induced by the spatial correlations in the data while excluding any effects coming from deviations of the temperature distribution from a Gaussian one.

To ensure the randomness of the set of Fourier phases we performed a rank-ordered remapping of the phases onto a set of uniformly distributed ones followed by an inverse Fourier transformation. These two preprocessing steps only have marginal influence to the maps. The main effect is that the outliers in the temperature distribution are removed. Due to the large number of temperature values (and phases) we did not find any significant dependence of the specific Gaussian (uniform) realization used for remapping of the temperatures (phases). The resulting maps may already be considered as a surrogate map and we named it zeroth order surrogate map. The first and second order surrogate maps are obtained as follows:

We first generate a first order surrogate map, in which any phase correlations for the scales, which are not of interest, are randomized. This is achieved by a random shuffle of the phases ϕ_{lm} for $l \notin \Delta l = [l_{min}, l_{max}], 0 < m \leq l$ and by performing an inverse Fourier transformation.

In a second step, N ($N = 500$ throughout this study) realizations of second order surrogate maps are generated for the first order surrogate map, in which the remaining phases ϕ_{lm} with $l \in \Delta l, 0 < m \leq l$ are shuffled, while the already randomized phases for the scales, which are not under consideration, are preserved. Note that the Gaussian properties of the maps, which are given by $|a_{lm}|$, are *exactly* preserved in all surrogate maps.

So far, we have applied the method of surrogates only to the l-range $\Delta l = [2, 20]$. In this chapter we will repeat the investigations for this l-interval but using newer CMB maps. Furthermore, we extend the analysis to smaller scales. Namely, we consider three more l-intervals $\Delta l = [20, 60]$, $\Delta l = [60, 120]$ and $\Delta l = [120, 300]$. The choice of 60 as l_{min} and l_{max} is somewhat arbitrary, whereas the $l_{min} = 120$ and $l_{max} = 300$ for the last l-interval was selected in such a way that the first peak in the power spectrum is covered. Going to even higher l's doesn't make much sense, because the ILC7 map is smoothed to 1 degree FWHM. Some other maps which we included in our study—especially NILC5—are not smoothed and we could in principle go to higher l's. But to allow for a consistent comparison of the results obtained with the different observed and simulated input maps we restrict ourselves to only investigate l-intervals up to $l_{max} = 300$ in this study.

Besides this two-step procedure aiming at a dedicated scale-dependent search of non-Gaussianity, we also test for non-Gaussianity using surrogate maps without specifying certain scales. In this case there are no scales, which are not of interest, and the first step in the surrogate map making procedure becomes dispensable. The zeroth order surrogate map is to be considered here as first order surrogate and the second order surrogates are generated by shuffling all phases with $0 < m \leq l$ for all available l's, i.e. in our case $\Delta l = [2, 1024]$.

Finally, for calculating scaling indices to test for higher order correlations the surrogate maps were degraded to $N_{side} = 256$ and residual monopole and dipole contributions were subtracted. The statistical comparison of the two classes of surrogates will reveal, whether possible HOCs on certain scales have left traces in the first order surrogate maps, which were then deleted in the second order surrogates. Before the results of such a comparison of the surrogate maps are shown in detail, we review the formalism of scaling indices.

6.4 Weighted Scaling Indices and Test Statistics

As test statistics for detecting and assessing possible scale-dependent non-Gaussianities in the CMB data weighted scaling indices are calculated [39, 40]. The basic ideas of the scaling index method (SIM) stem from the calculation of the dimensions of attractors in nonlinear time series analysis [45]. Scaling indices essentially represent one way to estimate the local scaling properties of a point set in an arbitrary d-dimensional embedding space. The technique offers the possibility of revealing local structural characteristics of a given point distribution. Thus, point-like, string-like and sheet-like structures can be discriminated from each other and from a random background. The alignment of e.g. string-like structures can be detected by using a proper metric for calculating the distances between the points [46, 47].

Besides the countless applications in time series analysis the use of scaling indices has been extended to the field of image processing for texture discrimination [48] and feature extraction [46, 49] tasks. Following further this line we performed several non-Gaussianity studies of the CMB based on WMAP data using scaling indices in recent years [24, 26, 40, 41].

Let us review the formalism for calculating this test statistic for assessing HOCs:

In general, the SIM is a mapping that calculates for every point $\vec{p}_i$, $i = 1, \ldots, N_{pix}$ of a point set P a single value, which depends on the spatial position of $\vec{p}_i$ relative to the group of other nearby points, in which the point under consideration is embedded in. Before we go into the details of assessing the local scaling properties, let us first of all outline the steps of generating a point set P out of observational CMB-data. To be able to apply the SIM on the spherical CMB data, we have to transform the pixelised sky S with its pixels at positions (θ_i, ϕ_i), $i = 1, ..., N_{pix}$, on the unit sphere to a point-distribution in an artificial embedding space. One way to achieve this is by transforming each temperature value $T(\theta_i, \phi_i)$ to a radial jitter around a sphere of radius R at the position of the pixel centre (θ_i, ϕ_i). Formally, the three-dimensional position vector of the point $\vec{p}_i$ reads as

$$x_i = (R + dR)\cos(\phi_i)\sin(\theta_i) \tag{6.1}$$

$$y_i = (R + dR)\sin(\phi_i)\sin(\theta_i) \tag{6.2}$$

$$z_i = (R + dR)\cos(\theta_i) \tag{6.3}$$

with

$$dR = a\left(\frac{T(\theta_i, \phi_i) - \langle T \rangle}{\sigma_T}\right). \tag{6.4}$$

Hereby, R denotes the radius of the sphere while a describes an adjustment parameter. The mean temperature and its standard deviation are characterised by $\langle T \rangle$ and σ_T, respectively. By the use of the normalisation we obtain for dR zero mean and a standard deviation of a. Both R and a should be chosen properly to ensure a high sensitivity of the SIM with respect to the temperature fluctuations at a certain spatial scale. For the analysis of WMAP-like CMB data, it turned out that this requirement is provided using $R = 2$ for the radius of the sphere and coupling the adjustment parameter a to the value of the below introduced scaling range parameter r via $a = r$ [24]. Now that we obtained our point set P, we can apply the SIM. For every point $\vec{p}_i$ we calculate the local weighted cumulative point distribution which is defined as

$$\rho(\vec{p}_i, r) = \sum_{j=1}^{N_{pix}} s_r(d(\vec{p}_i, \vec{p}_j)) \tag{6.5}$$

with r describing the scaling range, while $s_r(\bullet)$ and $d(\bullet)$ denote a shaping function and a distance measure, respectively. The scaling index $\alpha(\vec{p}_i, r)$ is then defined as the logarithmic derivative of $\rho(\vec{p}_i, r)$ with respect to r:

$$\alpha(\vec{p}_i, r) = \frac{\partial \log \rho(\vec{p}_i, r)}{\partial \log r}. \tag{6.6}$$

As mentioned above, $s_r(\bullet)$ and $d(\bullet)$ can in general be chosen arbitrarily. For our analysis we use a quadratic gaussian shaping function $s_r(x) = e^{-\left(\frac{x}{r}\right)^2}$ and an isotropic euclidian norm $d(\vec{p}_i, \vec{p}_j) = \|\vec{p}_i - \vec{p}_j\|$ as distance measure. With this specific choice of $s_r(\bullet)$ and $d(\bullet)$ we obtain the following analytic formula for the scaling indices

$$\alpha(\vec{p}_i, r) = \frac{\sum_{j=1}^{N_{pix}} 2\left(\frac{d_{ij}}{r}\right)e^{-\left(\frac{d_{ij}}{r}\right)^2}}{\sum_{j=1}^{N_{pix}} e^{-\left(\frac{d_{ij}}{r}\right)^2}}, \tag{6.7}$$

where we used the abbreviation $d_{ij} := d(\vec{p}_i, \vec{p}_j)$. As becomes obvious from Eq. (6.7), the calculation of scaling indices depends on the scale parameter r. Therefore, we can investigate the structural configuration in the underlying CMB-map in a scale-dependent manner. For our analysis, we use the ten scaling range parameters $r_k = 0.025, 0.05, ..., 0.25$, $k = 1, 2, ...10$, which (roughly) correspond to sensitive l-ranges from $\Delta l = [83; 387]$, $\Delta l = [41; 193]$, ..., $\Delta l = [8; 39]$ [26].

In order to quantify the degree of agreement between the surrogates of different orders with respect to their signatures left in distribution of scaling indices, we calculate the mean

$$\langle \alpha(r_k) \rangle = \frac{1}{N_p} \sum_{i=1}^{N_p} \alpha(\vec{p}_i, r_k) \tag{6.8}$$

and the standard deviation

$$\sigma_{\alpha(r_k)} = \left(\frac{1}{N_p - 1} \sum_{i=1}^{N_p} (\alpha(\vec{p}_i, r_k) - \langle \alpha(r_k) \rangle)^2 \right)^{1/2} \tag{6.9}$$

of the scaling indices α_i derived from N_p considered pixels for the different scaling ranges r_k. N_p becomes the number of all pixels N_{pix} for a full sky analysis. To investigate possible spatial variations of signatures of NG and to be able to measure asymmetries we also consider the moments as derived from the pixels belonging to rotated hemispheres. In these cases the number N_p of the pixels halves and their positions defined by the corresponding ϕ- and θ-intervals vary according to the part of the sky being considered. Furthermore, we combine these two test statistics by using χ^2 statistics. There is an ongoing discussion, whether a diagonal χ^2 statistic or the ordinary χ^2 statistic, which takes into account correlations among the different random variables through the covariance matrix is the better suited measure. On the one hand it is of course important to take into account correlations among the test statistics, on the other hand it has been argued [18] that the calculation of the inverse covariance matrix may become numerically unstable when the correlations among the variables are strong making the ordinary χ^2 statistic sensitive to fluctuations rather than to absolute deviations. Being aware of this we calculated both χ^2 statistics, namely the scale dependent diagonal χ^2 combining the mean and the standard deviation at a given scale r_k, and the scale-independent χ^2 combining the mean or/and the standard deviation calculated at all scales $r_k, k = 1, \ldots, 10$ (see [26]).

Further, we calculate the corresponding ordinary χ^2 statistics, which is obtained by summing over the full inverse correlations matrix $\mathbf{C}^{-1}$. In general, this is expressed by the bilinear form

$$\chi^2 = (\vec{M} - \langle \vec{M} \rangle)^T \mathbf{C}^{-1} (\vec{M} - \langle \vec{M} \rangle), \tag{6.10}$$

where the test statistics to be combined are comprised in the vector $\vec{M}$ and $\mathbf{C}$ is obtained by cross correlating the elements of $\vec{M}$. Specifically, for obtaining the scale dependent $\chi^2_{\text{full},\langle\alpha(r_k)\rangle,\sigma_{\alpha(r_k)}}$ combining the mean and the standard deviation at a given scale r_k the vector $\vec{M}^T$ becomes $\vec{M}^T = (M_1, M_2)$ with $M_1 = \langle\alpha(r_k)\rangle$, $M_2 = \sigma_{\alpha(r_k)}$.

Similarly, the full scale-independent χ^2 statistics $\chi^2_{\text{full},\langle\alpha\rangle}$, $\chi^2_{\text{full},\sigma_\alpha}$ and $\chi^2_{\text{full},\langle\alpha\rangle,\sigma_\alpha}$ are derived from the vectors $\vec{M}^T$ consisting of $\vec{M}^T = (\langle\alpha(r_1)\rangle, \ldots, \langle\alpha(r_{10})\rangle)$, $\vec{M}^T = (\sigma_{\alpha(r_1)}, \ldots, \sigma_{\alpha(r_{10})})$ and $\vec{M}^T = (\langle\alpha(r_1)\rangle, \ldots, \langle\alpha(r_{10})\rangle, \sigma_{\alpha(r_1)}, \ldots, \sigma_{\alpha(r_{10})})$, respectively. For all our investigations we calculated both χ^2 statistics and found out that the results are only marginally dependent from the chosen χ^2 statistics. Thus, in the following we will only list explicit numbers for the full χ^2 statistics, if not stated otherwise, because this measure yielded overall slightly more conservative results.

6.5 Results

To test for NGs and asymmetries in the ILC7 map and the NILC5 map, we compare the different surrogate maps in the following way:

For each scale we calculate the mean $\langle\alpha(r_k)\rangle$ and standard deviation $\sigma_{\alpha(r_k)}$ of the map of scaling indices $\alpha(\theta, \phi; r_k)$ of the full sky and a set of 768 rotated hemispheres. The northern pole of the different hemispheres is located at every pixel centre of the full sky with $N_{side} = 8$ in the HEALpix[2] [50] pixelisation scheme. The differences of the two classes of surrogates are quantified by the σ-normalized deviation S

$$S(Y) = \frac{Y_{\text{surro1}} - \langle Y_{\text{surro2}}\rangle}{\sigma Y_{\text{surro2}}} \tag{6.11}$$

with, $Y = \langle\alpha(r_k)\rangle, \sigma_{\alpha(r_k)}, \chi^2$. Every hemisphere of the set of 768 hemispheres delivers one deviation value S, which is then plotted on a sky map at that pixel position where the z-axis of the rotated hemisphere pierces the sky. Figure 6.1 shows the deviations S for the mean value $S(\langle\alpha(r_k)\rangle), k = 2, 6, 10$ for the ILC7 map as derived from the comparison of the different classes of surrogates for the scale-independent surrogate test and for the four selected l-ranges. The following striking features become immediately obvious:

First, various deviations representing features of non-Gaussianity and asymmetries can be found in the S-maps for the ILC7 map. These features can nearly exactly be reproduced when the NILC5 map is taken as input map (results not shown).

Second, we find for the scale-independent surrogate test (first row in Fig. 6.1) large isotropic deviations for the scaling indices calculated for the smallest scale shown in the figure. The negative values for S indicate that the mean of the scaling indices for the first order surrogate is smaller than for the second order surrogate maps. This systematic trend can be interpreted such that there's more structure detected in the first order surrogate than in the second order surrogate maps. Obviously, the random

---

[2] http://healpix.jpl.nasa.gov/

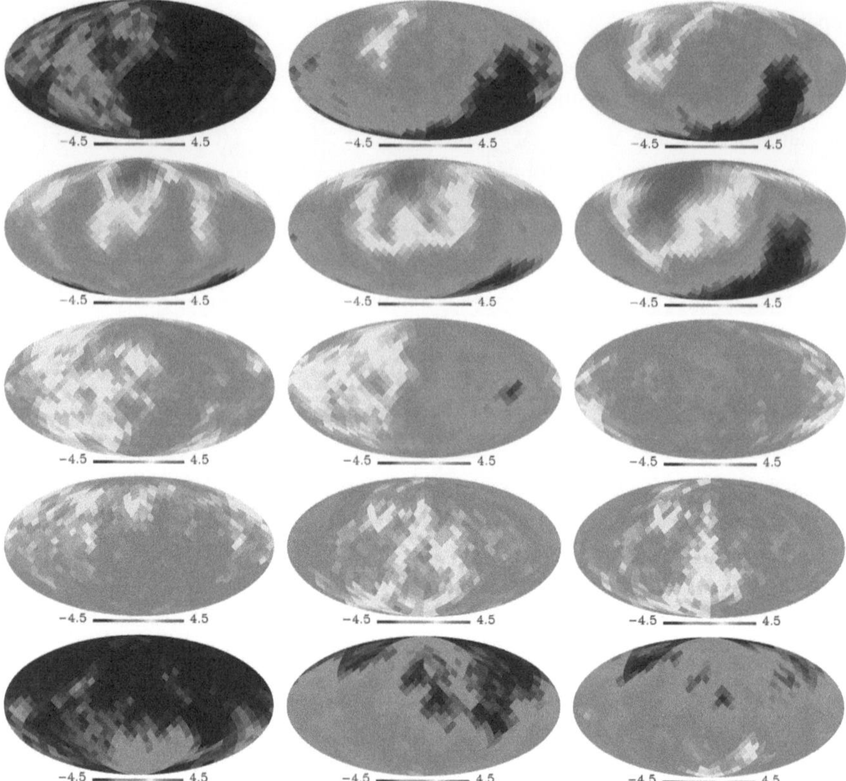

Fig. 6.1 Deviations $S(\langle\alpha(r_k)\rangle)$ of the rotated hemispheres for three scales r_k, $k = 2, 6, 10$ (from *left* to *right*) for the ILC7 map and for (from *top* to *bottom*) the shuffling intervals $\Delta l = [2, 1024]$, $\Delta l = [2, 20]$, $\Delta l = [20, 60]$, $\Delta l = [60, 120]$ and $\Delta l = [120, 300]$. The expected correspondence between the shuffling range Δl and the scales r_k of the scale-dependent higher order statistics $\langle\alpha(r_k)\rangle$, for which the largest deviations are detected, becomes apparent. While the ecliptic hemispherical asymmetries for $\Delta l = [2, 20]$ are most pronounced for the largest scaling range r_{10} (*second row*), the deviation S becomes largest for r_2 when shuffling the phases of the smallest scales $\Delta l = [120, 300]$ (*last row*)

shuffle of all phases has destroyed a significant amount of structural information at small scales in the maps.

Third, for the scale-dependent analysis we obtain for the largest scales ($\Delta l = [2, 20]$) highly significant signatures for non-Gaussianities and ecliptic hemispherical asymmetries at the largest r-values (second row in Fig. 6.1). These results are perfectly consistent with those obtained for the WMAP 5 yr ILC map and the foreground removed maps generated by [51] on the basis of the WMAP 3 year data (see [41]). The only difference between this study and our previous one is that we now obtain higher absolute values for S ranging now from $-4.00 < S < 3.72$ for the ILC7 map and $-4.36 < S < 4.50$ for the NILC5 map as compared to $-3.87 < S < 3.51$

for the WMAP 5 yr ILC map. Thus, the cleaner the map becomes due to better signal-to-noise ratio and/or improved map making techniques the higher the significances of the detected anomalies, which suggests that the signal is of intrinsic CMB origin.

Fourth, we also find for the smallest considered scales ($\Delta l = [150, 300]$) large isotropic deviations for the scaling indices calculated for a small scaling range r very similar to those observed for the scale-independent test.

Fifth, we do not observe very significant anomalies for the two other bands ($\Delta l = [20, 60]$ and $\Delta l = [60, 120]$) being considered in this study. Thus, the results obtained for the scale independent surrogate test can clearly be interpreted as a super-position of the signals identified in the two l-bands covering the largest ($\Delta l = [2, 20]$) and smallest $\Delta l = [120, 300]$) scales. Let us investigate the observed anomalies in more details. We begin with a closer look at the most significant deviations. Figure 6.2 shows the probability densities derived for the full sky and for (rotated) hemispheres for the scaling indices at the largest scaling range r_{10} for the first and second order surrogates for the l-interval $\Delta l = [2, 20]$. We recognize the systematic shift of the whole density distribution towards higher values for the upper hemisphere and to lower values for the lower hemisphere. As these two effects cancel each other for the full sky, we do no longer see significant differences in the probability densities in this case. Since the densities as a whole are shifted, the significant differences between first and second order surrogates found for the moments cannot be attributed to some salient localizable features leading to an excess (e.g. second peak) at very low or high values in otherwise very similar $P(\alpha)$-densities. Rather, the shift to higher (lower) values for the upper (lower) hemisphere must be interpreted as a global trend indicating that the first order surrogate map has less (more) structure than the respective set of second order surrogates. The seemingly counterintuitive result for the upper hemisphere is on the other hand consistent with a linear hemispherical structure analysis by means of a power spectrum analysis, where also a lack of power in the northern hemisphere and thus a pronounced hemispherical asymmetry was detected [19, 29]. However, it has to be emphasised that the effects contained in the power spectrum are—by construction—exactly preserved in both classes of surrogates, so that the scaling indices measure effects that can solely be induced by HOCs thus being of a new, namely non-Gaussian, nature. Interestingly though, the linear and nonlinear hemispherical asymmetries seem to be correlated with each other.

Figure 6.3 is very similar to Fig. 6.2 and shows the probability densities for the scaling indices calculated for the second smallest scaling range r_2 for the first and second order surrogates for the l-interval $\Delta l = [120, 300]$. The systematic shift towards smaller values for the first order surrogate for both hemispheres and thus for the full sky is visible. It is interesting to note that all densities derived from the ILC7 and NILC5 map differ significantly from each other. These differences can be attributed to e.g. the smoothing of the ILC7 map. However, the systematic differences between first and second order surrogates induced by the phase manipulations prevailed in all cases—irrespective of the input map.

The results for the deviations $|S(r)|$ for the full sky and rotated upper and lower hemisphere are shown for all considered l-ranges and all scales r in Fig. 6.4. The corresponding values for r_2 and r_{10} are listed in the Tables 6.1 and 6.2. In Table 6.3

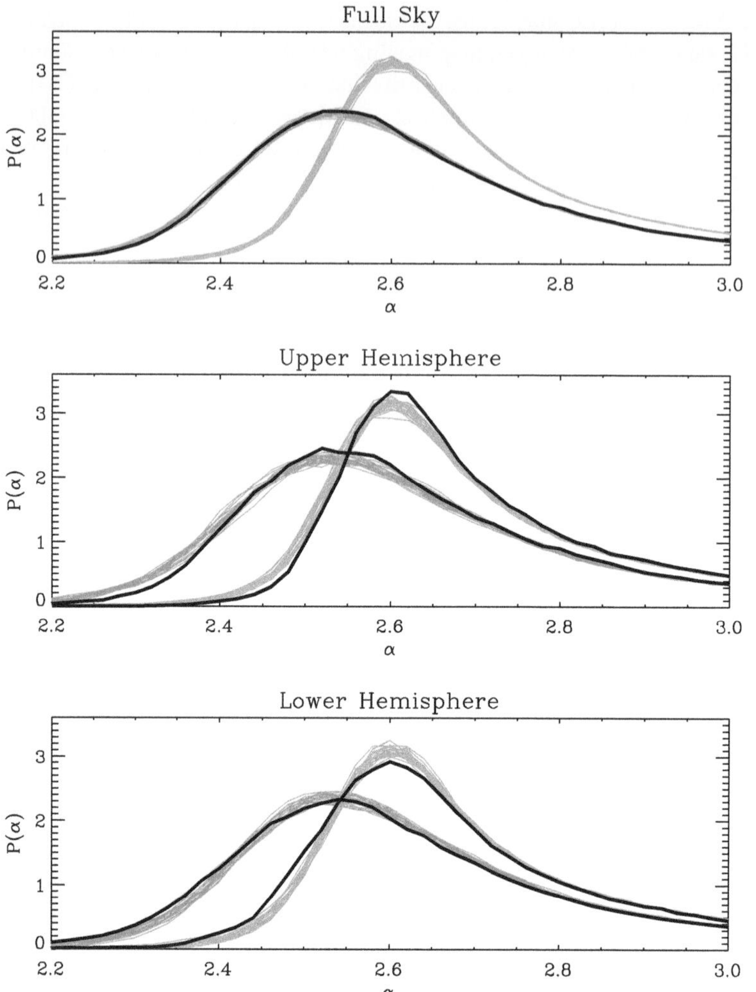

Fig. 6.2 Probability density $P(\alpha)$ of the first and second order surrogates for the scaling indices calculated for the largest scaling range r_{10} and for the l-interval $\Delta l = [2, 20]$. *Yellow (green) curves* denote the densities for 20 realizations of second order surrogates derived from the ILC7 (NILC5) map. The *black lines* are the corresponding first order surrogates. The reference frame for defining the upper and lower hemispheres is chosen such that the difference $\Delta S = S_{up} - S_{low}$ becomes maximal for $\langle \alpha \rangle$ of the respective map and respective scale r

we further summarize the results for the scale-independent χ^2-measures $\chi^2_{\langle \alpha \rangle}$, $\chi^2_{\sigma_\alpha}$ and $\chi^2_{\langle \alpha \rangle, \sigma_\alpha}$.

The main results which were already briefly discussed on the basis of Fig. 6.1 become much more apparent when interpreting Fig. 6.4 and Tables 6.1–6.3. We find stable $3.7 - 12\sigma$ deviations for all r-values for $S(\langle \alpha(r_k) \rangle)$ and the scale-independent

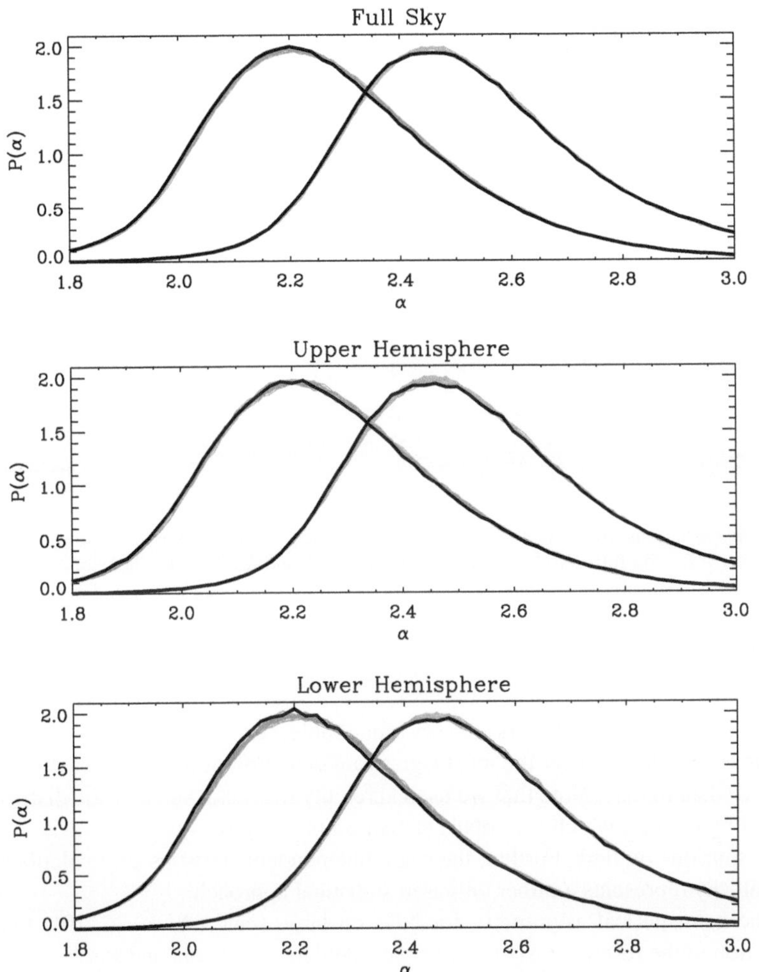

Fig. 6.3 Same as Fig. 6.2 but the second smallest scaling range r_2 and the l-interval $\Delta l = [120, 300]$

surrogate test when considering the full sky. This yields S-values of $S(\langle\alpha(r_2)\rangle) = 7.73$ (ILC7 map) and $S(\langle\alpha(r_2)\rangle) = 11.06$ (NILC5 map) for the scaling indices calculated for the small value r_2 and $S(\langle\alpha(r_{10})\rangle) = 3.75$ (ILC7 map) and $S(\langle\alpha(r_{10})\rangle) = 5.77$ (NILC5 map) for the largest radius r_{10}. This stable r-independent effect leads to very high values of the deviations S for the scale-independent χ^2-statistics $S(\chi^2_{\langle\alpha\rangle})$, where we find $S(\chi^2_{\langle\alpha\rangle}) = 5.73$ (ILC7 map) and $S(\chi^2_{\langle\alpha\rangle}) = 27.93$ (NILC5 map). It is interesting to compare these results with those obtained for the diagonal χ^2-statistics. In this case we find $S(\chi^2_{\langle\alpha\rangle}) = 57.32$ (ILC7 map) and $S(\chi^2_{\langle\alpha\rangle}) = 119.16$ (NILC5 map), which is up to an order of magnitude larger than the values for the

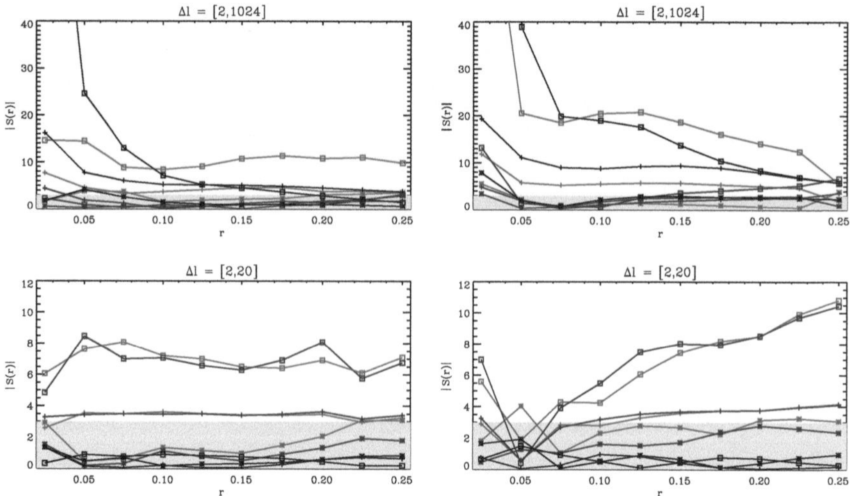

Fig. 6.4 Deviations $|S(r)|$ for the ILC7 (*left*) and NILC5 (*right*) map as a function of the scale parameter r for the full sky (*black*) and the *upper* (*red*) and *lower* (*blue*) hemisphere. The *plus* signs denote the results for the mean $\langle \alpha(r_k) \rangle$, the *star*-signs for the standard deviation $\sigma_{\alpha(r_k)}$ and the *boxes* for the χ^2-combination of $\langle \alpha(r_k) \rangle$ and $\sigma_{\alpha(r_k)}$. The *shaded region* indicates the 3σ significance interval

full χ^2-statistics. These results are very remarkable, since they represent—to the best of our knowledge—by far the most significant detection of non-Gaussianities in the WMAP data to date. Note that we used here only the mean value of the distribution of scaling indices, which is a robust statistics not being sensitive to contributions of some spurious outliers. Further, the scale-independent statistics $\chi^2_{\langle \alpha \rangle}$ calculated for the full sky represents a rather unbiased statistical approach.

The hemisperical asymmetry for NGs on large scale ($\Delta l = [2, 20]$) finds its reflection in the results of $S(r)$. While we calculate significant and stable deviations S for the upper and lower hemispheres separately (red and blue lines) in Fig. 6.4, the results for the full sky (black lines) are not significant, because the deviations detected in the two hemispheres are complementary and thus cancel each other. Therefore, we obtain only for the hemispheres high values for S ranging from $S = 3.24$ up to $S = 7.10$ ($S = 4.11$ up to $S = 10.82$) for the ILC7 (NILC5) map when considering the statistics derived from the scaling indices for the largest scales r_{10} and $S = 4.01$ up to $S = 9.76$ for the scale-independent χ^2-statistics.

For the smallest scales considered so far ($\Delta l = [120, 300]$) we also find significant deviations from non-Gaussianity being much more isotropic and naturally more pronounced at smaller scaling ranges $r < 0.15$. Thus, we obtain $S = 6.97$ (ILC7 map) and $S = 5.30$ (NILC5 map) for $S(\langle \alpha(r_2) \rangle)$ considering the full sky. For the scale-independent χ^2-statistics the most significant signatures of NGs are detected for the respective upper hemispheres ranging from $S = 5.16$ to $S = 10.53$. To test whether all these signatures are of intrinsic cosmic origin or more likely due to

Table 6.1 Deviations S and empirical probabilities p of the mean, standard deviation and their χ^2-combination as derived for the scaling indices at the second smallest scale r_2

Δl	Full sky ($S/\%$)	Upper hemisphere ($S/\%$)	Lower hemisphere ($S/\%$)
$\langle \alpha(r_2) \rangle$			
[2, 1024]	7.73/> 99.8	4.53/>99.8	1.87/96.0
[2, 20]	0.14/56.6	3.54/>99.8	3.44/>99.8
[20, 60]	0.88/80.6	1.84/96.4	1.08/85.2
[60, 120]	0.26/60.4	0.32/64.8	0.64/71.6
[120, 300]	6.97/>99.8	5.36/>99.8	0.92/83.0
$\sigma_{\alpha(r_2)}$			
[2, 1024]	4.16/>99.8	3.77/>99.8	0.25/61.8
[2, 20]	0.48/69.2	0.48/69.8	0.19/58.0
[20, 60]	1.70/95.2	3.18/>99.8	1.02/84.8
[60, 120]	0.88/80.0	2.35/98.8	1.25/88.2
[120, 300]	3.54/>99.8	1.03/83.4	3.69/>99.8
$\chi^2_{\langle\alpha(r_2)\rangle,\sigma_{\alpha(r_2)}}$			
[2, 1024]	24.55/>99.8	14.44/>99.8	0.94/84.4
[2, 20]	0.90/85.2	7.67/>99.8	8.47/99.8
[20, 60]	0.82/83.4	4.03/99.2	0.31/50.4
[60, 120]	0.51/61.4	3.63/98.6	1.00/85.2
[120, 300]	19.62/>99.8	17.17/>99.8	4.15/99.2

The results of the ILC7 map are shown for the different l-bands as well as for the full sky and the upper and lower hemispheres. Corresponding to the small scale r_2 the largest values for S are calculated for small scale non-Gaussianities in the l-range [120, 300] and for the scale-independent NGs, where the phases of all l's ($\Delta l = [2, 1024]$) are included

foregrounds or systematics induced by e.g. asymmetric beams or map making, we performed the same surrogate and scaling indices analysis for the five additional maps described in Sect. 6.2. Figures 6.5 and 6.6 show the significance maps for the two l-ranges $\Delta l = [2, 20]$ and $\Delta l = [120, 300]$, for which we found the most pronounced signatures in the ILC7 and NILC5 map. For the large scale NGs we find essentially the same results for the UILC7 map. The difference map, shows some signs of NGs and asymmetries, especially for large r-values. A closer look reveals, however, that both the numerator and denominator in the equation for S are an order of magnitude smaller than the values obtained for the ILC7 (NILC5) maps. Thus the signal of the difference map can be considered to be subdominant. And even if it were not subdominant, the signal coming from the residuals would rather diminish the signal in the ILC map than increase its significance, because the foreground signal is spatially anticorrelated with the CMB-signal. Both the asymmetric beam map and the simulated coadded VW-map do not show any significant signature for NGs and asymmetries. Finally, the simulated ILC map does show some signs of (galactic) north–south asymmetries which become smaller and therefore insignificant for increasing r, where we find the largest signal in the CMB maps.

For the small scale NGs ($\Delta l = [120, 300]$) we also find that the UILC7-map yields similar results as the ILC7 and NILC5 map with smaller significance. Once again the

Table 6.2 Same as Table 6.1 but for the scaling indices at the largest scale r_{10}

Δl	Full sky ($S/\%$)	Upper hemisphere ($S/\%$)	Lower hemisphere ($S/\%$)
$\langle\alpha(r_{10})\rangle$			
[2, 1024]	3.75/>99.8	3.53/>99.8	1.72/95.4
[2, 20]	0.64/74.2	3.24/>99.8	3.41/>99.8
[20, 60]	0.67/ 74.2	1.41/91.6	2.04/98.0
[60, 120]	0.01/50.5	2.28/99.0	2.19/98.6
[120, 300]	2.45/99.4	3.58/>99.8	1.38/92.2
$\sigma_{\alpha(r_{10})}$			
[2, 1024]	0.66/74.4	3.60/>99.8	2.90/>99.8
[2, 20]	0.84/80.0	3.09/>99.8	1.79/96.4
[20, 60]	2.27/98.6	2.94/99.8	0.13/55.0
[60, 120]	0.77/79.0	1.63/94.6	0.47/67.6
[120, 300]	0.60/73.6	1.61/95.8	0.81/79.6
$\chi^2_{\langle\alpha(r_{10})\rangle,\sigma_{\alpha(r_{10})}}$			
[2, 1024]	1.46/90.4	9.83/>99.8	3.15/98.0
[2, 20]	0.21/54.8	7.10/>99.8	6.77/99.8
[20, 60]	2.74/97.2	5.27/99.6	0.29/73.6
[60, 120]	0.38/50.2	2.09/94.2	0.43/75.8
[120, 300]	0.26/57.2	2.23/96.2	0.19/60.4

The largest values for S are found for large scales non-Gaussianities in the l-range [2, 20]

Table 6.3 Same as Table 6.1 but for the scale-independent χ^2-statistics

Δl	Full sky ($S/\%$)	Upper hemisphere ($S/\%$)	Lower hemisphere ($S/\%$)
$\chi^2_{\langle\alpha\rangle}$			
[2, 1024]	5.73/>99.8	9.35/>99.8	0.33/55.2
[2, 20]	0.97/95.0	4.57/99.6	4.01/99.2
[20, 60]	1.81/94.2	2.57/97.4	2.42/97.0
[60, 120]	1.41/99.0	1.53/99.6	0.91/83.8
[120, 300]	3.17/92.8	10.53/>99.8	1.19/87.8
$\chi^2_{\sigma_\alpha}$			
[2, 1024]	5.50/>99.8	11.50/>99.8	0.66/79.6
[2, 20]	0.32/52.8	4.03/98.6	4.04/99.6
[20, 60]	2.15/95.8	4.00/99.8	2.18/96.4
[60, 120]	1.40 /98.2	3.26/99.4	2.01/95.6
[120, 300]	3.10/99.0	8.90/>99.8	1.90/95.8
$\chi^2_{\langle\alpha\rangle,\sigma_\alpha}$			
[2, 1024]	1.89/94.2	8.38/>99.8	3.03/98.8
[2, 20]	0.73/77.4	5.64/>99.8	6.01/99.8
[20, 60]	1.60/92.8	3.42/99.2	1.49/91.0
[60, 120]	0.26/52.4	2.15/96.6	0.53/75.6
[120, 300]	1.68/92.8	5.34/99.8	0.22/63.2

Also for this statistics the largest values for S are found for the largest $\Delta l = [2, 20]$ and smallest scales $\Delta l = [120, 300]$ and for the scale-independent NGs

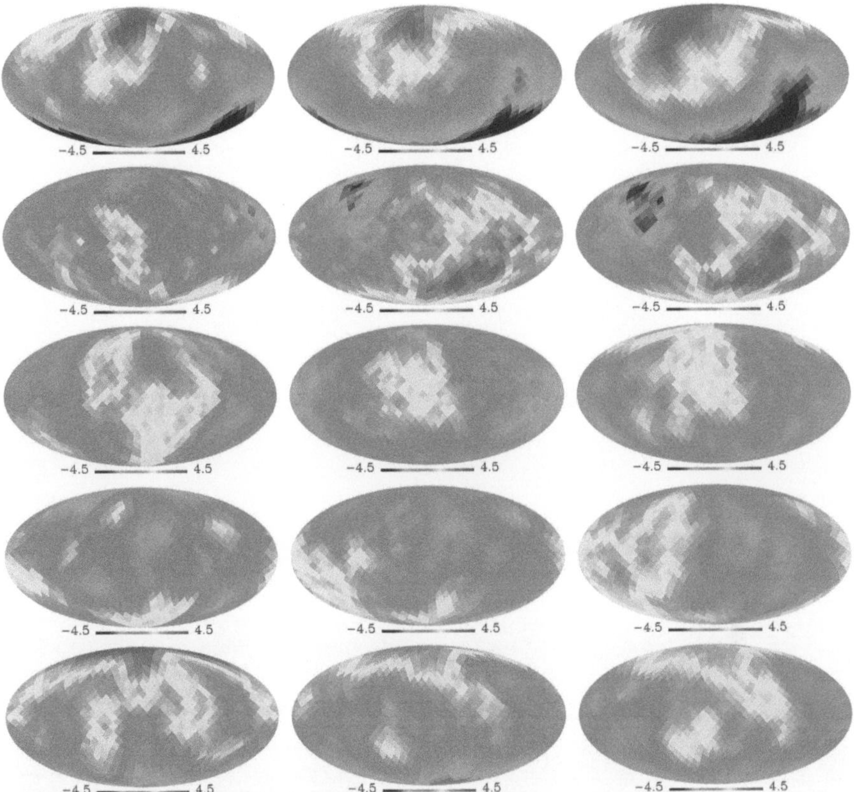

Fig. 6.5 Deviations $S(\langle\alpha(r_k)\rangle)$ for the three scales r_k, $k = 2, 6, 10$ (from *left* to *right*) for $\Delta l = [2, 20]$. The results are shown for (from *top* to *bottom*) the UILC7 map, the difference map 7yr ILC–6yr ILC map, the asymmetric beam map, the coadded V and W-band from a standard simulation and the simulated ILC-like map (for more detailed information about the different maps see text)

asymmetric beam map and the simulated coadded VW-map do not show significant signature for NGs and asymmetries. This is not the case for the simulated ILC map. Here, we find highly significant signatures for NGs and asymmetries, which show some similarities with significance patterns observed in the ILC7 (NILC5) map. Even much more striking features are detected in the difference map, where we find deviations as high as $|S| \approx 15$ forming a very peculiar pattern in the significance maps for all r. One of us (G.R.) named this pattern 'Eye of Sauron', which we think is a nice and adequate association. It is worth noticing that we found the same pattern when analyzing other difference maps, e.g. year 7–year 1 or year 2–year 1.

To better understand, where these features may come from we had a closer look at the zeroth, first and second order surrogate maps. It became immediately obvious that for the difference maps the fluctuations are systematically smaller in the regions in the galactic plane used for the ILC-map making than in the rest of the sky. This effect persists in the first order surrogate map and is only destroyed in the second

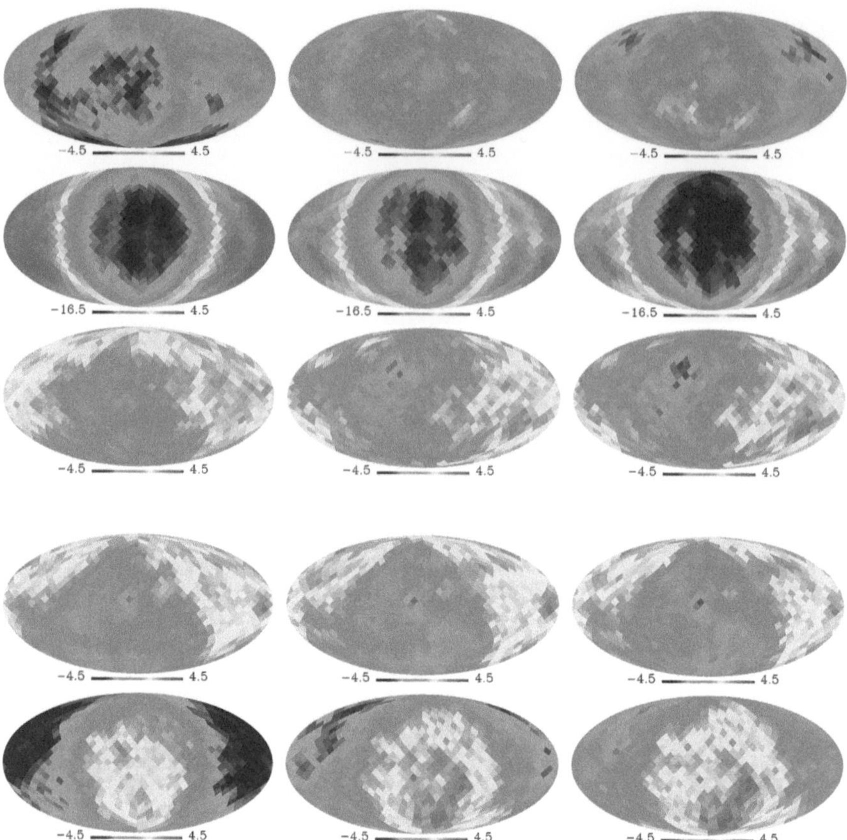

Fig. 6.6 Same as Fig. 6.5 but for $\Delta l = [120, 300]$. Note that the scale for the color coding has significantly changed for the difference map (*second row*)

order surrogates. This more (less) structure in first order surrogate map leads to lower (higher) values for the scaling indices, which can qualitatively explain the observed patterns in the significance maps.

A much more detailed study of these high l effects and their possible origins is part of our current work but is beyond the scope of this chapter. The results for the difference map shown here point, however, already towards a very interesting application of the surrogate technique. It may become a versatile tool to define criteria of the cleanness of maps in the sense of e.g. absence of artificially induced (scale-dependent) NGs in the map of the residual signal. Such a criterion may then in turn be implemented in the map making procedure so that ILC-like maps are not only minimizing the overall quadratic error in the map, but also e.g. the amount of unphysical NGs of the foregrounds.

6.6 Conclusions

To the best of our knowledge this work represents the first comprehensive study of scale-dependent non-Gaussianities in full sky CMB data as measured with the WMAP satellite. By applying the method of surrogate maps, which explicitly relies on the scale-dependent shuffling of Fourier phases while preserving all other properties of the map, we find highly significant signatures of non-Gaussianities for very large scales and for the l-interval covering the first peak in the power spectrum. In fact, our analyses yield by far the most significant evidence of non-Gaussianities in the CMB data to date. Thus, it is no longer the question whether there are phase correlations in the WMAP data. It is rather to be figured out what the origin of these scale-dependent non-Gaussian signatures is. The checks on systematics we performed so far revealed that no clear candidate can be found to explain the low-l signal, which we take to be cosmological at high significance. These findings would strongly disagree with predictions of isotropic cosmologies with single field slow roll inflation.

The picture is not that clear for the signatures found at smaller scales, i.e. at higher l's. In this case we found that NGs can also easily be induced by the ILC map making procedure so that it is difficult to disentangle possible intrinsic anomalies from effects induced by the preprocessing of the data. More tests are required to further pin down the origin of the detected high l anomalies and to probably uncover yet unknown systematics being responsible for the low l anomalies. Another way of ruling out effects of unknown systematics is to perform an independent observation preferably via a different instrument as we are now able to do with the Planck satellite.

In any case our study has shown that the method of surrogates in conjunction with sensitive higher order statistics offers the potential to become an important tool not only for the detection of scale-dependent non-Gaussianity but also for the assessment of possibly induced artefacts leading to NGs in the residual map which in turn may have important consequences for the map making procedures.

References

1. E. Komatsu, J. Dunkley, M.R. Nolta, C.L. Bennett, B. Gold, G. Hinshaw, N. Jarosik, D. Larson, M. Limon, L. Page, D.N. Spergel, M. Halpern, R.S. Hill, A. Kogut, S.S. Meyer, G.S. Tucker, J.L. Weiland, E. Wollack, E.L. Wright, ApJ Suppl. Ser. **180**, 330 (2009)
2. E. Komatsu, K.M. Smith, J. Dunkley, C.L. Bennett, B. Gold, G. Hinshaw, N. Jarosik, D. Larson, M.R. Nolta, L. Page, D.N. Spergel, M. Halpern, R.S. Hill, A. Kogut, M. Limon, S.S. Meyer, N. Odegard, G.S. Tucker, J.L. Weiland, E. Wollack, E.L. Wright, ApJS **192**, 18 (2011)
3. E. Komatsu, N. Afshordi, N. Bartolo, D. Baumann, J.R. Bond, E.I. Buchbinder, C.T. Byrnes, X. Chen, D.J.H. Chung, A. Cooray, P. Creminelli, N. Dalal, O. Dore, R. Easther, A.V. Frolov, K.M. Górski, M.G. Jackson, J. Khoury, W.H. Kinney, L. Kofman, K. Koyama, L. Leblond, J.-L. Lehners, J.E. Lidsey, M. Liguori, E.A. Lim, A. Linde, D.H. Lyth, J. Maldacena, S. Matarrese, L. McAllister, P. McDonald, S. Mukohyama, B. Ovrut, H.V. Peiris, C. Raeth, A. Riotto, Y. Rodriguez, M. Sasaki, R. Scoccimarro, D. Seery, E. Sefusatti, U. Seljak, L. Senatore, S. Shandera, E.P.S. Shellard, E. Silverstein, A. Slosar, K.M. Smith, A.A. Starobinsky, P.J. Steinhardt,

F. Takahashi, M. Tegmark, A.J. Tolley, L. Verde, B.D. Wandelt, D. Wands, S. Weinberg, M. Wyman, A.P.S. Yadav, M. Zaldarriaga, Astro2010, The Astronomy and Astrophysics Decadal Survey. Science White Papers **158** (2009)

4. A.H. Guth, Phys. Rev. D **23**, 347 (1981)
5. A. Linde, Phys. Lett. **108B**, 389 (1982)
6. A. Albrecht, P. Steinhardt, Phys. Rev. Lett. **48**, 1220 (1982)
7. A.H. Linde, V. Mukhanov, Phys. Rev. D **56**, R535 (1997)
8. P.J.E. Peebles, ApJ Lett. **438**, L1 (1997)
9. F. Bernardeau, J.-P. Uzan, Phys. Rev. D **66**, 103506 (2002)
10. V. Acquaviva, N. Bartolo, S. Matarrese, A. Riotto, Nucl. Phys. B **667**, 119 (2003)
11. C. Armendariz-Picon, T. Damour, V. Mukhanov, Phys. Lett. B **458**, 209 (1999)
12. J. Garriga, V. Mukhanov, Phys. Lett. B **458**, 219 (1999)
13. X. Chen, Phys. Rev. D **72**, 123518 (2005)
14. M. LoVerde, A. Miller, S. Shandera, L. Verde, JCAP **4**, 14 (2008)
15. E.I. Buchbinder, J. Khoury, B.A. Ovrut, PRL **100**, 171302 (2008)
16. J.-L. Lehners, P.J. Steinhardt, Phys. Rev. D **78**, 023506 (2008)
17. C.-G. Park, MNRAS **349**, 313 (2004)
18. H.K. Eriksen, D.I. Novikov, P.B. Lilje, A.J. Banday, K.M. Górski, ApJ **612**, 64 (2004)
19. F.K. Hansen, A.J. Banday, K.M. Górski, MNRAS **354**, 641 (2004)
20. P. Vielva, E. Martínes-González, R.B. Barreiro, J.L. Sanz, L. Cayón, ApJ **609**, 22 (2004)
21. H.K. Eriksen, A.J. Banday, K.M. Gorski, P.B. Lilje, ApJ **622**, 58 (2005)
22. H.K. Eriksen, A.J. Banday, K.M. Gorski, F.K. Hansen, P.B. Lilje, ApJ Lett. **660**, L81 (2007)
23. A. Oliveira-Costa, M. Tegmark, M. Zaldarriaga, A. Hamilton, Phys. Rev. D **69**, 063516 (2004)
24. C. Räth, P. Schuecker, A.J. Banday, MNRAS **380**, 466 (2007)
25. J.D. McEwen, M.P. Hobson, A.N. Lasenby, D.J. Mortlock, MNRAS **388**, 659 (2008)
26. G. Rossmanith, C. Räth, A.J. Banday, G. Morfill, MNRAS **399**, 1921 (2009)
27. C.J. Copi, D. Huterer, D.J. Schwarz, G.D. Starkman, MNRAS **399**, 295 (2009)
28. C.J. Copi, D. Huterer, D.J. Schwarz, G.D. Starkman, Adv. Astr. **847541** (2010)
29. F.K. Hansen, A.J. Banday, K.M. Górski, H.K. Eriksen, P.B. Lilije, ApJ **704**, 1448 (2009)
30. A. Yoho, F. Ferrer, G.D. Starkman, arXiv1005.5389 (2011)
31. R. Zhang, D. Huterer, Astropart. Phys. **33**, 69 (2010)
32. C.L. Bennett, R.S. Hill, G. Hinshaw, D. Larson, K.M. Smith, J. Dunkley, B. Gold, M. Halpern, N. Jarosik, A. Kogut, E. Komatsu, M. Limon, S.S. Meyer, M.R. Nolta, N. Odegard, L. Page, D.N. Spergel, G.S. Tucker, J.L. Weiland, E. Wollack, E.L. Wright, ApJS **192**, 17 (2011)
33. L.-Y. Chiang, P.D. Naselsky, O.V. Verkhodanov, M.J. Way, ApJ Lett. **590**, L65 (2003)
34. P. Coles, P. Dineen, J. Earl, D. Wright, MNRAS **350**, 989 (2004)
35. P.D. Naselsky, L.-Y. Chiang, P. Olesen, I. Novikov, Phys. Rev. D **72**, 063512 (2005)
36. L.-Y. Chiang, P.D. Naselsky, P. Coles, ApJ **664**, 8 (2007)
37. J. Theiler, S. Eubank, A. Longtin, B. Galdrikian, J.D. Farmer, Phys. D **58**, 77 (1992)
38. A. Bunde, J. Kropp, H.-J. Schellnhuber, *The Science of Disasters: Climate Disruptions, Heart Attacks, and Market Crashes* (Springer, Berlin, 2002)
39. C. Räth, W. Bunk, M.B. Huber, G.E. Morfill, J. Retzla, P. Schuecker, MNRAS **337**, 413 (2002)
40. C. Räth, P. Schuecker, MNRAS **344**, 115 (2003)
41. C. Räth, G. Morfill, G. Rossmanith, A.J. Banday, K.M. Gorski, PRL **102**, 131301 (2009)
42. B. Gold, N. Odegard, J.L. Weiland, R.S. Hill, A. Kogut, C.L. Bennett, G. Hinshaw, X. Chen, J. Dunkley, M. Halpern, N. Jarosik, E. Komatsu, D. Larson, M. Limon, S.S. Meyer, M.R. Nolta, L. Page, K.M. Smith, D.N. Spergel, G.S. Tucker, E. Wollack, E.L. Wright, ApJS **192**, 15 (2011)
43. J. Delabrouille, J.-F. Cardoso, M. Le Jeune, M. Betoule, G. Fay, F. Guilloux, A&A **493**, 835 (2009)
44. I.K. Wehus, L. Ackerman, H.-K. Eriksen, N.E. Groeneboom, ApJ **707**, 343 (2009)
45. P. Grassberger, I. Procaccia, PRL **50**, 346 (1983)
46. C. Räth, R. Monetti, J. Bauer, I. Sidorenko, D. Müller, M. Matsuura, E.-M. Lochmüller, P. Zysset, F. Eckstein, New J. Phys. **10**, 125010 (2008)

47. K.R. Sütterlin, A. Wysocki, A.V. Ivlev, C. Räth, H.M. Thomas, M. Rubin-Zuzic, W.J. Goedheer, V.E. Fortov, A.M. Lipaev, V.I. Molotkov, O.F. Petrov, G.E. Morfill, H. Löwen, PRL **102**, 085003 (2009)
48. C. Räth, G.E. Morfill, JOSA A **14**, 3208 (1997)
49. F. Jamitzky, R.W. Stark, W. Bunk, S. Thalhammer, C. Räth, T. Aschenbrenner, G.E. Morfill, W.M. Heckl, Ultramicroscopy **86**, 241 (2001)
50. K.M. Górski, E. Hivon, A.J. Banday, B.D. Wandelt, F.K. Hansen, M. Reinecke, M. Bartelmann, ApJ **622**, 759 (2005)
51. M. Tegmark, A. Oliveira-Costa, A. Hamilton, Phys. Rev. D **68**, 123523 (2003)

Chapter 7
Applying the Surrogate Approach to Incomplete Skies

Introduction.—The search for primordial non-Gaussianities in the Cosmic Microwave Background (CMB) is one of the most important yet challenging tasks in modern cosmology. Any convincing detection of intrinsic non-Gaussianities as well as their characteristics and scaling behaviour would directly support or reject different models of inflation, and therefore affect a fundamental part of the standard cosmological model.

The currently still favoured inflationary model is single-field slow-roll inflation [1–3], which should result in (nearly) Gaussian and isotropic temperature fluctuations of the CMB. However, preferred directions and other kinds of asymmetries have been repeatedly detected [4–17], already questioning the simplest picture of inflation. It is under discussion, if these asymmetries are connected to foreground influences [18, 19], which appear particularly in the direction of the Galactic plane.

For investigations of CMB data sets, e.g. WMAP data, the analysis of Fourier phases has proven to be a useful method [20–23], since all potential higher order correlations, which directly point to non-Gaussianities, are contained in the phases and the correlations among them. The method of surrogate maps with shuffled Fourier phases [11, 12] represents one way of analysing the phases. Originally, this idea stems from the field of time series analysis [24–27] and describes the construction of data sets, so-called surrogates, which are similar to the original, except for a few modified characteristics. The validation of these characteristics in the original data can then be tested by comparing them to the set of surrogates with appropriate measures. The method used in [11, 12] tests the hypothesis that the coefficients $a_{lm} = |a_{\ell m}| e^{i\phi_{lm}}$ of the Fourier transform of the temperature values $T(\theta, \phi)$ have independent and uniform distributed phases $\phi_{\ell m} \in [-\pi, \pi]$ calculated for the complete sphere S. The phases of the original map are shuffled, which can be done within some previously chosen interval of interest, $\Delta \ell = [\ell_1, \ell_2]$, or simply for the complete range

Original publication: G. Rossmanith, C. Räth, A. J. Banday, H. Modest, K. M. Górski, G. E. Morfill, *Probing non-Gaussianities in the CMB on an incomplete sky using surrogates*, PRL, submitted (2011). Figure 7.3 was added to this version.

G. Rossmanith, *Non-linear Data Analysis on the Sphere*, Springer Theses, DOI: 10.1007/978-3-319-00309-2_7, © Springer International Publishing Switzerland 2013

$\Delta \ell = [2, \ell_{max}]$ for some given ℓ_{max}. Every realisation of this shuffling results in a new set of $a_{\ell m}$'s, which then represents (after transforming back) one surrogate map. Note that every surrogate still has by construction exactly the same power spectrum as the original map. If the original map contained any phase correlations, these are now destroyed due to the shuffling. Thus, any detected differences between the original and a set of surrogate maps reveals higher order correlations and therefore deviations from Gaussianity.

One major problem in CMB analyses is the treatment of the Galactic plane, which strongly influences the microwave signal. It is possible to cut out the foreground affected regions [28], but this procedure itself can affect the subsequent analyses as well. When applying a sky cut, orthonormality of the spherical harmonics no longer holds on this new incomplete sky, which leads to a coupling of the $a_{\ell m}$'s, making a naive phase shuffling impossible. However, one can transform the spherical harmonics into a new set of harmonics, which forms an orthonormal basis on the incomplete sky [29–31], where phase manipulation can then take place again.

The problem of incomplete data also occurs in time series analysis by means of surrogates. Here, gaps can be overcome e.g. by the use of simulated annealing [32, 33]. Still, the quality of surrogates constructed with this method seems to be questionable, since it is not ensured that no phase correlations are induced.

In this *Letter*, we combine the cut sky methods with phase shuffling, thus enabling investigations by means of surrogates on an incomplete sky. Our method can also be extended for the usage on incomplete data sets in general.

Methods.—On a complete sphere S, an orthonormal basis is given by the spherical harmonics $Y_{\ell m}(s)$ with $\ell \geq 0$, $-\ell \leq m \leq \ell$ and $s \in S$. Let the number of harmonics be limited by some given $\ell_{max} \in \mathbb{N}^+$. Now, for any map

$$f(s) = \sum_{\ell,m}^{\ell_{max}} a_{\ell m} Y_{\ell m}(s), \ \forall s \in S$$

with $a_{\ell m} \in \mathbb{C}$, and for any new incomplete sky S^{cut}, we want to know the corresponding $a_{\ell m}^{cut}$ and $Y_{\ell m}^{cut}$ for representing the map on the remaining regions of the sphere:

$$f(s) = \sum_{\ell,m}^{\ell_{max}} a_{\ell m}^{cut} Y_{\ell m}^{cut}(s), \ \forall s \in S^{cut},$$

with $Y_{\ell m}^{cut}$ being orthonormal on S^{cut} and thus $a_{\ell m}^{cut}$ being unique. For real valued spherical harmonics, this was performed in [29] and [30], and later on extended in [31]. The methods presented there can be easily adopted to the complex valued spherical harmonics as well.

At first, we define the vectors

$$Y(s) := [Y_{0,0}(s), Y_{1,0}(s), Y_{1,1}(s), \ldots, Y_{\ell_{max},\ell_{max}}(s)]^T,$$

$$a \quad := [a_{0,0}, a_{1,0}, a_{1,1}, \ldots, a_{\ell_{max},\ell_{max}}]^T$$

containing all harmonics and coefficients with $m \geq 0$, respectively, of the given map on the complete sphere. Both have a lenght of $i_{max} := (\ell_{max} + 1)(\ell_{max} + 2)/2$. Analogously, we define $Y^{cut}(s)$ and a^{cut}. Our objective is to determine two transformation matrices $B_1, B_2 \in \mathbb{C}^{i_{max} \times i_{max}}$, that fulfil the following equations:

$$Y^{cut}(s) = B_1 Y(s) \tag{7.1}$$

$$a^{cut} = B_2 a \tag{7.2}$$

To identify them, we need to define the coupling matrix

$$C := \int_R Y(s)Y^*(s)d\Omega$$

as well as analogously its counterpart C^{cut}, with R being a given region on the sphere. Hereby, Y^* denotes the hermitian transposed of Y. When working with a pixelised sky, one uses a sum over the pixels of R instead of the integral. For $R = S^{cut}$, an orthonormal set of harmonics $Y^{cut}_{\ell m}$ needs to fulfill the condition $C^{cut} = I_{i_{max}}$, with $I_{i_{max}}$ being the unit matrix of size i_{max}. We can use Eq. (7.1) on C^{cut} to change this condition to $B_1 C B_1^* = I_{i_{max}}$. It is possible to apply different matrix decompositions to obtain $C = AA^*$ with $A \in \mathbb{C}^{i_{max} \times i_{max}}$. Consequently, the above equation now reads as $(B_1 A)(B_1 A)^* = I_{i_{max}}$ and offers the simple solution $B_1 = A^{-1}$.

For the evaluation of B_2, let us recall that the coefficient vector a can be expressed by

$$a = \int_S \overline{Y}(s)f(s)d\Omega$$

or, respectively,

$$a^{cut} = \int_{S^{cut}} \overline{Y}^{cut}(s)f(s)d\Omega.$$

Inserting (7.1) and the expression $f(s) = a^T Y(s)$ into the latter leads to $a^{cut} = \overline{B_1}C^T a$. Now we use the above matrix decomposition again and obtain $a^{cut} = \overline{B_1}(AA^*)^T a = A^T a$. Thus, it follows that $B_2 = A^T$.

To obtain the $a^{cut}_{\ell m}$ and $Y^{cut}_{\ell m}$ with $m < 0$, we make use of the following equations, that hold for full sky and that we assume to be valid also on incomplete skies:

$$Y^{cut}_{\ell,-m} = (-1)^{|m|}\overline{Y}^{cut}_{\ell m}$$

and

$$a^{cut}_{\ell,-m} = (-1)^{|m|}\overline{a}^{cut}_{\ell m}.$$

For the sky cuts and ℓ-ranges used throughout this *Letter*, the cut sky harmonics were tested and confirmed to be orthogonal.

For constructing $C = AA^*$ as above, one can make use of different matrix decomposition methods. Since the coupling matrix C is hermitian and can be treated as positive definite for low ℓ by construction, a Cholesky decomposition is applicable. This is the easiest and fastest way, although numerical problems only allow usage for lower ℓ_{max} [31]. Another possibility is the eigendecomposition (ED): We obtain $C = VWV^*$, with the columns of V containing the eigenvectors, and W being diagonal and containing the eigenvalues of C. Because of the properties of the coupling matrix, these values are real and positive, allowing therefore a simple decomposition of W by taking the square root of every element, $W = W^{1/2}(W^*)^{1/2}$. Thus, we obtain $A = VW^{1/2}$. Since C is hermitian, the ED is formally similar to a singular value decomposition (SVD), which is also applied in this *Letter*, with the eigenvalues corresponding to the singular values. For both the ED and the SVD we apply a householder transformation similar to [31] to make A lower triangular. For the Cholesky decomposition, this is already the case by definition. Thus, due to Eq. (7.2), it is ensured that the mono- and dipole contributions of the underlying maps—often considered as non-cosmological—are kept separate from the $\ell \geq 2$ modes.

With the help of the new cut sky harmonics $Y_{\ell m}^{cut}$, we can now generate the surrogates on a cut sky S^{cut} as well. Similar to above, we shuffle the phases $\phi_{\ell m}^{cut}$ of the cut sky coefficients $a_{\ell m}^{cut}$, which is in this work performed for the full cut sky range $\Delta \ell^{cut} = [2, \ell_{max}]$. We obtain new sets of $a_{\ell m}^{cut}$'s, which are transformed back to pixel space to form the cut sky surrogate (CSS) maps. As we did in the case of a complete sphere, we now search for deviations between the original data as well as its surrogates. However, one has to take care about the above mentioned properties. While the uniform distribution still holds for $\phi_{\ell m}^{cut}$, the single phases in the sets are no longer independent from each other due to Eq. (7.2). In other words, the cut sky transformation induces phase correlations to the underlying map. To account for these systematic effects, we create for each of the input maps 20 full sky surrogate (FSS) maps as explained above, with $\ell_{max} = 1024$ and by shuffling the phases within $\Delta \ell = [2, 1024]$. By comparing the results of the surrogate analysis for an input map and its FSS, we evade systematically induced phase correlations and search for additional signatures possibly contained in the phases.

In general, the comparison of the original data and its surrogate maps can be accomplished with any higher order statistics. In this *Letter*, we chose the *scaling index method* (SIM) [10, 12] as well as *Minkowski functionals* [4, 34] as test statistics.

The SIM is a local measure that is able to detect structural characteristics of a given data set by estimating its local scaling properties. Briefly, the temperature anisotropies $T(\theta, \phi)$ are transformed to variations in radial direction around the sphere, therefore leading to a point distribution $\vec{p}_i, i = 1, \ldots, N_{pix}$, in three-dimensional space. Then, the weighted cumulative point distribution $\rho(\vec{p}_i, r)$ is calculated for every point $\vec{p}_i$ and a freely chosen scaling parameter r. Since we will only investigate the large scales in this *Letter*, we choose the free parameter r to be $r_{10} = 0.25$, which is appropriate for these scales [10]. Eventually, the scaling indices are obtained by calculating the logarithmic derivative of $\rho(\vec{p}_i, r)$ with respect to r.

The three Minkowski functionals measure the behaviour of a given map with respect to different threshold values ν. The fraction of the sky where the temperature value is larger than ν is denoted as the excursion set $R(\nu)$, its smooth boundary is identified by $\partial R(\nu)$, and da and dl describe the surface element of $R(\nu)$ and the line element of $\partial R(\nu)$, respectively. Then, we can define the three Minkowski functionals as

$$M_{area}(\nu) = \int_{R(\nu)} da$$

$$M_{perim}(\nu) = \int_{\partial R(\nu)} dl$$

$$M_{euler}(\nu) = \int_{\partial R(\nu)} dl\kappa,$$

with κ being the geodesic curvature of $\partial R(\nu)$. For more details, we refer to [4, 34]. Eventually, we sum up over all thresholds with the help of the appropriate cut sky surrogates by means of a χ^2-measure,

$$\chi_\bullet^2 = \sum_{\nu} \left[\left(M_\bullet^{map}(\nu) - \langle M_\bullet^{CSS}(\nu) \rangle \right) / \sigma_{M_\bullet^{CSS}(\nu)} \right]^2$$

for $M_{area}(\nu)$, $M_{perim}(\nu)$ and $M_{euler}(\nu)$, respectively.

The results for the different maps of both the scaling indices and the Minkowski functionals are then evaluated in terms of *rotated hemispheres*: For 768 different angles we rotate the underlying maps and calculate the σ-normalised deviations

$$S_1(Y) = \left(Y^{map} - \langle Y^{CSS} \rangle \right) / \sigma_{Y^{CSS}}$$

of the pixels included in the new upper hemisphere between the input map and its cut sky surrogates, by means of the measure Y. In our case, $Y = \langle \alpha \rangle, \sigma_\alpha, \chi^2_{area}, \chi^2_{perim}, \chi^2_{euler}$, with $\langle \alpha \rangle$ and σ_α being the mean and the standard deviation of the scaling index response $\alpha(s)$, respectively. The result is then shown as colour-coded pixel, whose centre is pierced by the z-axis of the respective rotated reference frame (see [10–12]). To separate traces of possibly intrinsic phase correlations from those induced by the transition to incomplete sky, we calculate the statistics

$$S_2(Y) = \left(S_1^{data}(Y) - \langle S_1^{FSS}(Y) \rangle \right) / \sigma_{S_1^{FSS}(Y)}$$

for comparing the results of $S_1(Y)$ for the original and the full sky surrogate maps.

To investigate possible deviations from statistical isotropy, we introduce an asymmetry statistics $\Delta S_2(Y)$. For the scaling indices, we define the difference in Y between each pair of opposite hemispheres as

$$\Delta S_2(Y) = \left| S_2^{up}(Y) - S_2^{low}(Y) \right|$$

for $Y = \langle \alpha \rangle, \sigma_\alpha$. This statistics is appropriate for the SIM, since the sign of the deviations is preserved in $S_2(Y)$. This is not the case for the quadratic measure χ^2 used on the Minkowski functionals. Therefore, we have to include the difference between opposite hemispheres already in the χ^2-measure by computing

$$\Delta \chi_\bullet^2 = \sum_\nu \left[\left(\Delta M_\bullet^{map}(\nu) - \langle \Delta M_\bullet^{CSS}(\nu) \rangle \right) / \sigma_{\Delta M_\bullet^{CSS}(\nu)} \right]^2 ,$$

with

$$\Delta M_\bullet(\nu) = M_\bullet^{up}(\nu) - M_\bullet^{low}(\nu) .$$

We can then define the asymmetry statistics for the Minkowski functionals as

$$\Delta S_2(Y) = S_2(\Delta Y)$$

for $Y = \chi_{area}^2, \chi_{perim}^2, \chi_{euler}^2$.

We construct the cut sky harmonics for three different central latitude sky cuts, that remove $|b| < 10°, 20°$ and $30°$ of latitude in the centre of the maps. While the smallest cut ($|b| < 10°$) removes already a large amount of the highly foreground affected regions but retains nearly all non-affected regions, the largest ($|b| < 30°$) excludes almost the entire Galactic plane, with only minor point sources remaining. We choose an upper bound of $\ell_{max} = 20$ and set $a_{\ell m} = 0$ for $\ell > \ell_{max}$. To check for consistency with [11, 12], we applied the cut sky formalism also to the complete sphere with no points excluded. To compare the different matrix decomposition methods, all three approaches (Cholesky, ED, SVD) were applied. For every sky cut, the phases of the coefficients $a_{\ell m}^{cut}$ were shuffled to generate $N = 100$ cut sky surrogates for each input map. The same was done for the corresponding FSS maps. For computational reasons, the resolution of the input maps in the corresponding HEALPix scheme [35, 36] was chosen to be $N_{side} = 256$ for the scaling index analysis and $N_{side} = 64$ for the Minkowski functionals. By testing several subsets, we assured ourselves that the results are only marginally affected when choosing a lower resolution.

Validation.—To test the new approach, we generate a Gaussian simulation of the coadded VW-band of the WMAP satellite via a noise-weighted sum. The procedure is the same as in [10], but note that we now apply the more recent WMAP 7-year parameters. In addition, we applied the cut sky surrogate approach to another simulated Gaussian map, to which we added typical foreground residuals that are still present after the template cleaning of the WMAP data. Those residuals were computed by subtracting the WMAP ILC map from the full seven-year foreground reduced coadded VW-band. This is done to examine the impact of possible aliasing effects due to the chosen ℓ_{max}, that could be caused by strong foregrounds which are cut out.

For the clean simulated Gaussian map, the significances $S_2(\sigma_\alpha)$ calculated for the rotated hemispheres are illustrated in Fig. 7.1. The differences between the original

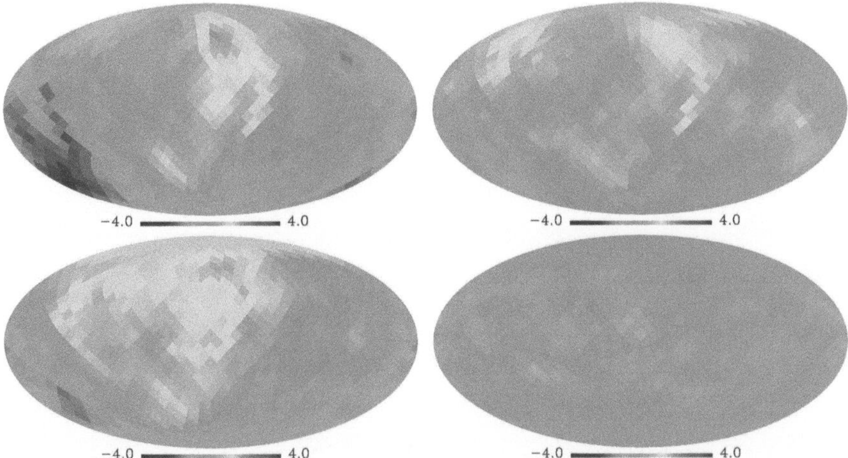

Fig. 7.1 The σ-normalised deviations $S_2(\sigma_\alpha)$ comparing a simulated map and its 20 full-sky surrogates for the complete sphere (*upper left*) and the three different central latitude sky cuts $|b| < 10°$ (*upper right*), $|b| < 20°$ (*lower left*), and $|b| < 30°$ (*lower right*), that were constructed by means of the singular value decomposition

and the FSS maps are insignificant ($S_2(Y) < 3$) for the complete sphere and all sky cuts. The same holds for the assembled map, except for the full sky, where phase correlations are obviously present. These results clearly demonstrate the practicality of the approach. The results of the assembled map show that the impact of aliasing effects, even when strong foregrounds are present in the Galactic plane, is negligible.

Only minimal differences were detected for the three used matrix decomposition methods, that are likely to be due to the random shuffle of the phases. When going to larger ℓ-ranges or more irregular sky cuts, this technical part of the investigation will become more important, especially for making the transition to the cut sky possible.

Application to WMAP data.—For the application of the cut sky method to observational data, we make use of two different maps, which are both linear combinations of the different frequency bands and based on the WMAP results [37, 38]: First, the 7-year Internal Linear Combination (ILC7) map provided by the WMAP team [28] and second the 5-year needlet based ILC map (NILC5) [39]. For both maps, the monopole and dipole were removed.

The significances $S_2(\sigma_\alpha)$ determined for the rotated hemispheres of the NILC5 map for the different sky cuts are shown in Fig. 7.2 while the findings for $\Delta S_2(Y)$ are illustrated in Fig. 7.4. When looking at the deviations $S_2(\langle\alpha\rangle)$ and $S_2(\sigma_\alpha)$ for the $|b| < 10°$ cut of both the ILC7 and NILC5 maps, we detect significant non-Gaussianities and an asymmetry. Both features were already found in corresponding full-sky analyses [11, 12]. The signal for $S_2(\langle\alpha\rangle)$ becomes more unarticulate for larger cuts. Still, the maximum difference $\Delta S_2(\langle\alpha\rangle)$ between opposite hemispheres remains significant. For $S_2(\sigma_\alpha)$, a significant asymmetry with a clear north-south

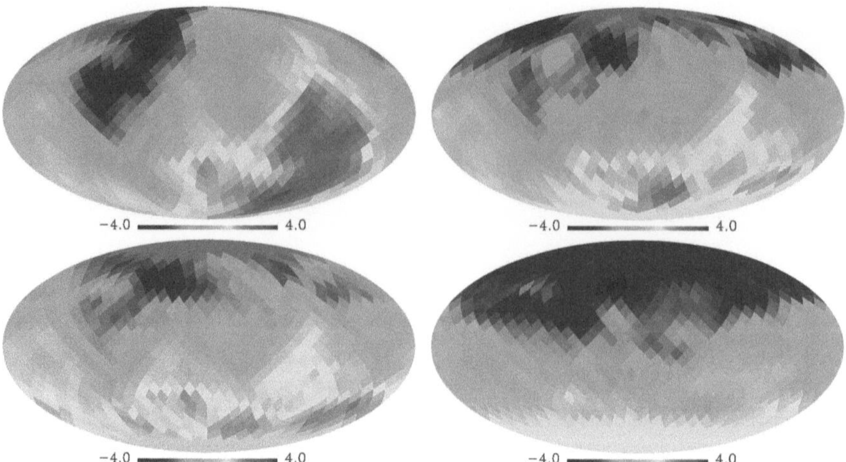

Fig. 7.2 Same as Fig. 7.1 but for the NILC5 map

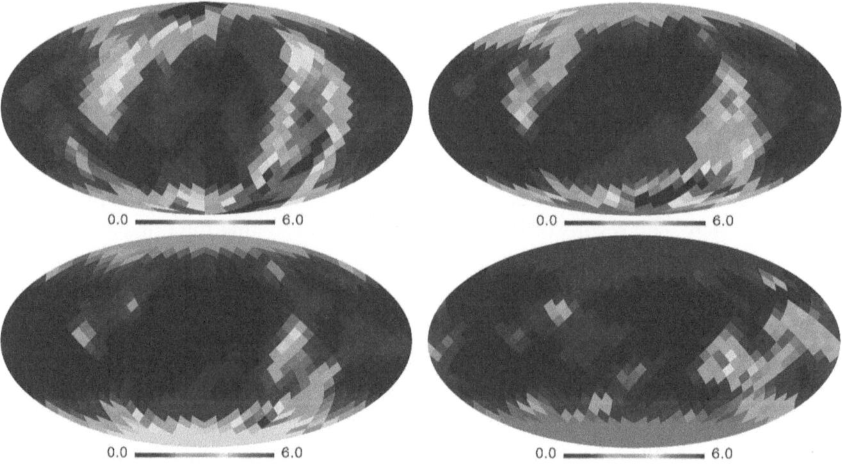

Fig. 7.3 Same as Fig. 7.2 but for $S_2(\chi^2_{area})$

direction persists for both the ILC7 and the NILC5 maps when excluding the Galactic plane, which is also reflected by a constant $\Delta S_2(\sigma_\alpha)$.

The Minkowski functionals show similar results: For $M_{area}(\nu)$, one detects similar deviations between the data sets and its full sky surrogates. This is illustrated for the NILC5 map in Fig. 7.3. Again, these findings are reflected in the asymmetry statistics $\Delta S_2(\chi^2_{area})$, as shown in Fig. 7.4. For $M_{perim}(\nu)$ and $M_{euler}(\nu)$, the results agree for the full sky, but get less definite for larger cuts, which is likely to be due to the limited amount of pixels.

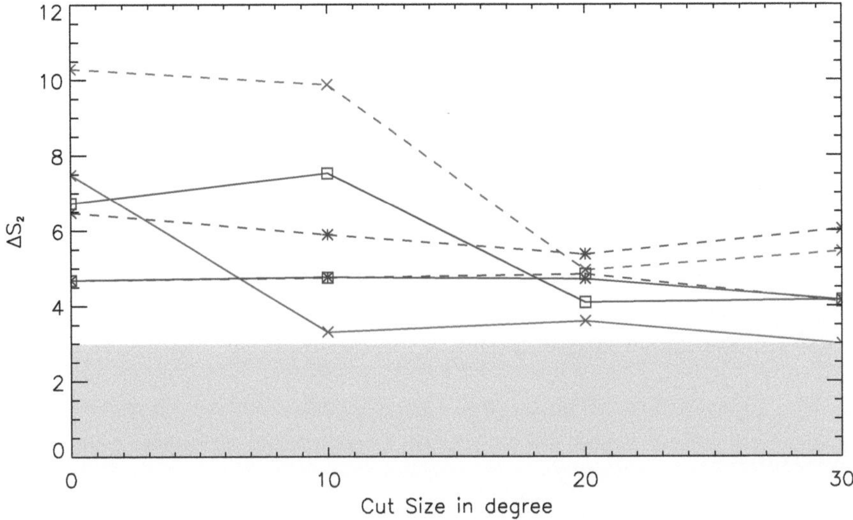

Fig. 7.4 The difference between the results of opposite hemispheres $\Delta S_2(Y)$, for the ILC7 (*solid*) and the NILC5 (*dashed*) maps. The reference frame for defining the upper and lower hemispheres is chosen such that $\Delta S_2(Y)$ becomes maximal. The *blue lines* with the *boxes* and the *star-signs* denote the results of $\langle \alpha \rangle$ and σ_α, respectively, while the *red lines* mark the results of χ^2_{area}

The results of the data clearly indicate that both the detected non-Gaussianity and asymmetry cannot mainly be attributed to foreground influences. In combination with the multitude of checks on systematics performed for the surrogates technique in [11, 12], one has to conclude that the signatures are of cosmological origin. This represents a strong violation of the Gaussian hypothesis and of statistical isotropy. Both assumptions are fundamental parts of single-field slow-roll inflation, which is therefore rejected at high significance by this analysis. In addition, due to the fact that all cuts remove a notable amount of pixels from the sphere (for the largest cut $|b| < 30°$ it is already half of the sky), the decreasing significance for the incomplete skies can at least in parts be explained by less input points, which leads to an increasing influence of noise and a lower effect of the intrinsic signal. This especially holds for the two Minkowski functionals $M_{perim}(\nu)$ and $M_{euler}(\nu)$, which examine complex pixel formations and thus need enough data points to produce statistically reliable results.

Conclusion.—We demonstrated the feasibility of generating surrogates by Fourier-based methods also for an incomplete data set. This was worked out for the case of a CMB analysis on an incomplete sphere. Three different constant latitude sky cuts were applied. For this purpose, three different cut sky transformations were calculated. We generated 100 cut sky surrogates for every input map, sky cut and matrix decomposition method, which were analysed by means of scaling indices and Minkowski functionals. To remove systematic effects, a second analysis compared the results of the original with the ones of 20 full sky surrogate maps for each of the input maps. For simulated maps, no anomalies could be detected. The findings for the ILC7 and the NILC5 maps show strong signatures of non-Gaussianities and pronounced asymmetries, which persist even when removing larger parts of the sky. This confirms that the influence of the Galactic plane is not responsible for these deviations from Gaussianity and isotropy. Together with former full-sky analyses, the results point towards a violation of statistical isotropy. Similar tests with the forthcoming PLANCK-data will yield more information about the origin of the detected anomalies.

Many of the results in this chapter have been obtained using HEALPix [35]. We acknowledge the use of LAMBDA. Support for LAMBDA is provided by the NASA Office of Space Science.

References

1. A.H. Guth, Phys. Rev. D **23**, 347 (1981)
2. A. Linde, Phys. Lett. **108B**, 389 (1982)
3. A. Albrecht, P. Steinhardt, Phys. Rev. Lett. **48**, 1220 (1982)
4. H.K. Eriksen, D.I. Novikov, P.B. Lilje, A.J. Banday, K.M. Górski, ApJ **612**, 64 (2004)
5. H.K. Eriksen, A.J. Banday, K.M. Gorski, F.K. Hansen, P.B. Lilje, ApJ Lett. **660**, L81 (2007)
6. F. K. Hansen, H. K. Eriksen, A. J. Banday, K. M. Gorski, P. B. Lilje, Cosm. Front. **379**, 16–23 (2007)
7. C. Dickinson, H.K. Eriksen, A.J. Banday, J.B. Jewell, K.M. Górski, G. Huey, C.R. Lawrence, I.J. O'Dwyer, B.D. Wandelt, Astrophys. J. **705**, 1607 (2009)
8. J. Hoftuft, H.K. Eriksen, A.J. Banday, K.M. Górski, F.K. Hansen, P.B. Lilje, ApJ **699**, 985 (2009)
9. F.K. Hansen, A.J. Banday, K.M. Górski, H.K. Eriksen, P.B. Lilije, ApJ **704**, 1448 (2009)
10. G. Rossmanith, C. Räth, A.J. Banday, G. Morfill, MNRAS **399**, 1921 (2009)
11. C. Räth, G. Morfill, G. Rossmanith, A.J. Banday, K.M. Gorski, PRL **102**, 131301 (2009)
12. C. Räth, A.J. Banday, G. Rossmanith, H. Modest, R. Sütterlin, K.M. Górski, J. Delabrouille, G.E. Morfill, Mon. Not. R. Astron. Soc. **415**, 2205 (2011)
13. P. Vielva, E. Martínes-González, M. Cruz, R.B. Barreiro, M. Tucci, MNRAS, arXiv:1002.4029v2
14. D. Pietrobon, P. Cabella, A. Balbi, R. Crittenden, G. de Gasperis, N. Vittorio, Mon. Not. R. Astron. Soc. Lett. **402**, L34 (2010)
15. F. Paci, A. Gruppuso, F. Finelli, P. Cabella, A. De Rosa, N. Mandolesi, P. Natoli, MNRAS **407**, 339 (2010)
16. L. Cayón, MNRAS **405**, 1084 (2010)
17. C.J. Copi, D. Huterer, D.J. Schwarz, G.D. Starkman, Adv. Astron. **847541** (2010)
18. C.J. Copi, D. Huterer, D.J. Schwarz, G.D. Starkman, MNRAS **399**, 295 (2009)
19. P.D. Naselsky, P.R. Christensen, P. Coles, O. Verkhodanov, D. Novikov, J. Kim, Astrophys. Bull. **65**, 101 (2010)
20. L.-Y. Chiang, P.D. Naselsky, O.V. Verkhodanov, M.J. Way, ApJ Lett. **590**, L65 (2003)
21. P. Coles, P. Dineen, J. Earl, D. Wright, MNRAS **350**, 989 (2004)
22. P.D. Naselsky, L.-Y. Chiang, P. Olesen, I. Novikov, Phys. Rev. D **72**, 063512 (2005)
23. L.-Y. Chiang, P.D. Naselsky, P. Coles, ApJ **664**, 8 (2007)
24. J. Theiler, S. Eubank, A. Longtin, B. Galdrikian, J.D. Farmer, Physica D **58**, 77 (1992)
25. T. Schreiber, A. Schmitz, Phys. Rev. Lett. **80**, 2105 (1998)
26. A. Bunde, J. Kropp, H.-J. Schellnhuber, *The Science of Disasters: Climate Disruptions, Heart Attacks, and Market Crashes* (Springer, Berlin, 2002)

27. M. Gliozzi, C. Räth, I.E. Papadakis, P. Reig, A&A **512**, A21 (2010)
28. B. Gold, N. Odegard, J.L. Weiland, R.S. Hill, A. Kogut, C.L. Bennett, G. Hinshaw, X. Chen, J. Dunkley, M. Halpern, N. Jarosik, E. Komatsu, D. Larson, M. Limon, S.S. Meyer, M.R. Nolta, L. Page, K.M. Smith, D.N. Spergel, G.S. Tucker, E. Wollack, E.L. Wright, ApJS **192**, 15 (2011)
29. K.M. Górski, ApJ Lett. **430**, L85 (1994)
30. K.M. Górski, ApJ Lett. **430**, L89 (1994)
31. D.J. Mortlock, A.D. Challinor, M.P. Hobson, MNRAS **330**, 405 (2002)
32. T. Schreiber, A. Schmitz, Phys. Rev. Lett. **79**, 1475 (1997)
33. T. Schreiber, A. Schmitz, Physica D **142**, 346 (2000)
34. K. Michielsen, H. De Raedt, Phys. Rep. **347**, 461 (2001)
35. K.M. Górski, E. Hivon, A.J. Banday, B.D. Wandelt, F.K. Hansen, M. Reinecke, M. Bartelmann, ApJ **622**, 759 (2005)
36. http://healpix.jpl.nasa.gov
37. N. Jarosik, C.L. Bennett, J. Dunkley, B. Gold, M.R. Greason, M. Halpern, R.S. Hill, G. Hinshaw, A. Kogut, E. Komatsu, D. Larson, M. Limon, S.S. Meyer, M.R. Nolta, N. Odegard, L. Page, K.M. Smith, D.N. Spergel, G.S. Tucker, J.L. Weiland, E. Wollack, E.L. Wright, ApJS **192**, 14 (2011)
38. http://lambda.gsfc.nasa.gov/
39. J. Delabrouille, J.-F. Cardoso, M. Le Jeune, M. Betoule, G. Fay, F. Guilloux, A&A **493**, 835 (2009)

Chapter 8
Conclusions

In this work, the five- and seven-year observations of the CMB by the WMAP satellite were analysed in detail by means of the scaling index method. The basic ideas for this method stem from the calculation of dimensions of strange attractors in nonlinear time series analysis. Scaling indices are able to identify and characterise the structural components of a given data set. One the one hand, the tests for deviations from the standard model were performed by comparing the data sets to simulations of Gaussian random fields mimicking the properties of the ΛCDM model, which represents the standard approach in CMB investigations. On the other hand, a novel approach, namely the method of surrogates, was developed in this work, which offers the possibility to analyse the CMB in a completely data-driven way. Here, the basic idea is the construction of surrogate data sets, which are generated by dedicated shuffling of the Fourier phases of the original data map.

Foregrounds, in particular present in the Galactic plane, lead to strong distortions of the CMB measurements. The best way to handle these distortions is to mask the respective regions, which causes in turn problems concerning the analysing procedure. First, it affects the evaluation of the scaling indices close to the mask. For this reason, a mask-filling method was developed in this work that prevents these boundary effects. Second, the orthonormality of the Fourier basis set is violated, which is required for the applicability of the surrogates approach. In this work, the method of surrogates was—for the first time—successfully combined with a basis transformation that creates a set of orthonormal Fourier functions on cut skies, thus enabling the surrogates approach on an incomplete sphere. This method was carefully tested to assess and then rule out the effects of systematics.

The different analyses of the WMAP data performed with the scaling index method lead to the following results:

- For the WMAP five-year data, an analysis by means of simulated CMB maps showed strong evidence for non-Gaussianity, detected an obvious asymmetry, and revealed several local features in the data, whereupon all results were in agreement with former investigations. All examined bands lead to consistent findings.

G. Rossmanith, *Non-linear Data Analysis on the Sphere*, Springer Theses, DOI: 10.1007/978-3-319-00309-2_8, © Springer International Publishing Switzerland 2013

- An investigation of different available five- and seven-year full-sky maps by means of surrogates—representing the first application of this method—lead to the by far most significant detection of non-Gaussianity to date. Detailed checks on systematic found no non-cosmological origin for these anomalies.
- The combination of surrogate approach and cut sky transformation was applied to data sets for the first time. The analysis identified once more highly significant non-Gaussianities and asymmetries, now even for incomplete sky coverage where the entire Galactic plane was removed. This confirms that the strong foreground effects present in the Galactic plane are not responsible for the deviations from Gaussianity and isotropy. In addition to scaling indices, this investigation was also performed with Minkowski functionals. Different techniques for the basis transformation were applied as well. Both statistics showed consistent results for all different basis transformation techniques.

Summarising all these results, the standard picture of single-field slow-roll inflation is strongly questioned. In addition, the findings point towards a violation of statistical isotropy in general. The techniques that were developed in this work are ready to be used on upcoming data sets. Independent and even more precise measurements, in particular with the current PLANCK satellite, will possibly reveal the true nature of the beginnings of our Universe.

Appendix
Simplifications of the Cut Sky Approach

In this appendix, two technical approaches are introduced, that can significantly simplify the transformation of the spherical to the cut sky harmonics from Sect. 2.1.3. The first method describes an optional way to calculate the coupling matrix C if a constant latitude cut is applied to the sphere. This method is numerically preferable to the usual direct computation, since for these type of cuts many components of C become trivial. The second method described here is the Householder transformation. With its help, we can modify the matrix A to be lower triangular. The advantage of such a matrix is that the potentially non-cosmological mono- and dipole are kept separate from the other modes during the transformation to the cut sky regime, and can therefore easily be removed. For the Cholesky decomposition, a Householder transformation is obsolete since in this case A is already lower triangular by definition.

Both the simplification of the constant latitude cuts as well as the Householder transformation described in this appendix are complex-valued extensions of the real-valued methods used in [1], and were applied throughout this work.

A.1 Constant Latitude Cuts

For the construction of the coupling matrix

$$C = \int_R Y(\vec{x})Y^*(\vec{x})d\Omega \tag{A.1}$$

in Sect. 2.1.3, one needs to define the remaining surface R of the sphere S. If this area is exclusively defined by one latitude interval of the form $\theta_1 \leq \theta \leq \theta_2$ or a combination of several of those intervals, but independent of the longitude φ, with $0 \leq \theta \leq \pi$, $0 \leq \varphi \leq 2\pi$, the applied sky cut is denoted constant latitude cut. In this case, only the components $C_{i(\ell,m),i(\ell',m')}$ with $m = m'$ are non-zero. This simplifies the computation of C significantly. In addition, C is real-valued for the constant latitude cuts.

A proof for both of these statements is given in the following, where the remaining surface is defined as $R = \{(\theta, \varphi)|\, \theta_1 \leq \theta \leq \theta_2\}$ for simplicity reasons.

G. Rossmanith, *Non-linear Data Analysis on the Sphere*, Springer Theses, 123
DOI: 10.1007/978-3-319-00309-2, © Springer International Publishing Switzerland 2013

Equation (A.1) in the component-wise form reads as

$$C_{i(\ell,m),i(\ell',m')} = \int_R Y_{i(\ell,m),i(\ell',m')}(\vec{x})\overline{Y}_{i(\ell,m),i(\ell',m')}(\vec{x})d\Omega$$

and can be rewritten by means of latitude and longitude as

$$C_{i(\ell,m),i(\ell',m')} = \int_0^{2\pi}\int_{\theta_1}^{\theta_2} Y_{i(\ell,m),i(\ell',m')}(\theta,\varphi)\overline{Y}_{i(\ell,m),i(\ell',m')}(\theta,\varphi)\sin\theta d\theta d\varphi .$$

By inserting the definition of the spherical harmonics (see Sect. 1.3.3), one can separate these two integrals, which leads to

$$C_{i(\ell,m),i(\ell',m')} = k\int_0^{2\pi} e^{im\varphi}e^{-im'\varphi}d\varphi \int_{\theta_1}^{\theta_2} P_{\ell m}(\cos\theta)P_{\ell'm'}(\cos\theta)\sin\theta d\theta ,$$

whereupon we made use of the definition

$$k := \frac{1}{2\pi}\sqrt{\frac{2\ell+1}{2}\frac{(\ell-m)!}{(\ell+m)!}}\sqrt{\frac{2\ell'+1}{2}\frac{(\ell'-m')!}{(\ell'+m')!}}.$$

Both the factor k as well as the second integral are real-valued. Therefore, it is sufficient to show that the first integral is zero for $m \neq m'$, and real-valued otherwise. This can be done by applying Euler's formula:

$$\int_0^{2\pi} e^{im\varphi}e^{-im'\varphi}d\varphi = \int_0^{2\pi} e^{i(m-m')\varphi}d\varphi$$

$$= \int_0^{2\pi} \cos((m-m')\varphi)\varphi d\varphi + i\int_0^{2\pi} \sin((m-m')\varphi)\varphi d\varphi$$

For $m \neq m'$, the term $(m-m')$ is some non-zero integer. Thus, one integrates over one or more complete oscillations of the sinus or cosinus function, respectively, which leads to the annihilation of the positive and negative parts. For $m = m'$, the first integral is 2π, and the second zero. In summary, one can write

$$C_{i(\ell,m),i(\ell',m')} = 2\pi k\,\delta_{m,m'}\int_{\theta_1}^{\theta_2} P_{\ell m}(\cos\theta)P_{\ell'm'}(\cos\theta)\sin\theta d\theta .$$

and therefore the above statements are proven.

A.2 The Householder Transformation

For the Cholesky decomposition, the matrix $A \in \mathbb{C}^{i_{max}\times i_{max}}$ from Sect. 2.1.3 is lower triangular. This is advantageous for two reasons: One the one hand, it slightly simplifies the computation of the cut sky harmonics $Y_{\ell m}^{cut}$ and its coefficients $a_{\ell m}^{cut}$. On the other hand, due to Eq. (2.6), the mono- and dipole have no influence on the other modes when the transformation to the cut sky regime is applied. Therefore, these potentially non-cosmological modes can easily be removed in the cut sky, too.

For the eigenvalue and singular value decompositions, A is not triangular in general. But it is possible to transform A into a lower triangular matrix A' by means of the Householder transformation (e.g. [2]). Since the transformation itself creates an upper triangular matrix, we need to focus on the transposed counterpart A^T in the following. Formally, the transformation of A^T into an upper triangular matrix $(A^T)'$ due to multiplication with an unitary matrix P reads as

$$(A^T)' = PA^T . \tag{A.2}$$

Both the matrices $(A^T)'$ and P_i have the same size as A^T. This procedure is iterative: In step i, the multiplication $(A^T)^{i+1} = P_i(A^T)^i$ shall set the last $(i_{max} - i)$ components of the ith column of the matrix $(A^T)^i$ to zero:

$$\begin{bmatrix} a'_{1,1} & & & & & \\ 0 & \ddots & & & & \\ \vdots & & a'_{i-1,i-1} & & & \\ & & 0 & a'_{i,i} & & \\ & & 0 & 0 & a'_{i+1,i+1} & \cdots \\ & & \vdots & \vdots & \vdots & \\ 0 & \cdots & 0 & 0 & a'_{i_{max},i+1} & \cdots \end{bmatrix}$$

$$= P_i \begin{bmatrix} a_{1,1} & & & & & \\ 0 & \ddots & & & & \\ \vdots & & a_{i-1,i-1} & & & \\ & & 0 & a_{i,i} & & \\ & & 0 & a_{i+1,i} & a_{i+1,i+1} & \cdots \\ & & \vdots & \vdots & \vdots & \\ 0 & \cdots & 0 & a_{i_{max},i} & a_{i_{max},i+1} & \cdots \end{bmatrix}$$

Here, $a_{i,i}$ denotes the components of $(A^T)^i$ and $a'_{i,i}$ the ones of $(A^T)^{i+1}$. The transformation can be accomplished by means of the matrix

$$P_i = \begin{bmatrix} I_{i-1} & 0 \\ 0 & p_i \end{bmatrix},$$

whereby I_{i-1} denotes the unit matrix of size $(i-1)$, and the matrix $p_i \in \mathbb{C}^{(i_{max}-i+1)\times(i_{max}-i+1)}$ is defined as

$$p_i := I_{i_{max}-i+1} - 2\frac{m_i m_i^T}{m_i^T m_i}\,,$$

with

$$(m_i)_j := \begin{cases} a_{i,i} + \sqrt{\sum_{k=i}^{i_{max}} a_{k,i}\overline{a}_{k,i}} & \text{for } j = 1, \\ a_{j,i} & \text{for } j > 1. \end{cases}$$

In summary, we obtain

$$P = P_1 P_2 \ldots P_{i_{max}-1}\,.$$

Due to the fact that P is unitary, Eq. (A.2) can be written as

$$A = A'U$$

with $U = \overline{P}^{-1}$ and $A' = (A^T)'^T$. The matrix U is unitary as well, and does therefore not affect the decomposition equation $C = AA^*$ from Sect. 2.1.3. Thus, one is able to make use of a triangular transformation matrix even in the case of an eigenvalue or singular value decomposition.

References

1. D.J. Mortlock, A.D. Challinor, M.P. Hobson, MNRAS **330**, 405 (2002)
2. W.H. Press, S.A. Teukolsky, W.T. Vetterling, B.P. Flannery, *Numerical Recipes: The Art of Scientific Computing*, 3rd edn (Cambridge University Press, Cambridge, 2007)

MIX
Papier aus verantwortungsvollen Quellen
Paper from responsible sources
FSC® C105338

If you have any concerns about our products,
you can contact us on
ProductSafety@springernature.com

In case Publisher is established outside the EU,
the EU authorized representative is:
**Springer Nature Customer Service Center GmbH
Europaplatz 3, 69115 Heidelberg, Germany**

Printed by Libri Plureos GmbH
in Hamburg, Germany